GCSE 9-1

HIGHER

MATHEMATICS

REVISION GUIDE

FOR ALL EXAM BOARDS

Stephen Doyle

Author Stephen Doyle
Editorial team Haremi Ltd
Series designers emc design ltd
Typesetting York Publishing Solutions Pvt. Ltd.
Illustrations York Publishing Solutions Pvt. Ltd.
App development Hannah Barnett, Phil Crothers and Haremi Ltd

Designed using Adobe InDesign
Published by Scholastic Education, an imprint of Scholastic Ltd, Book End, Range Road, Witney, Oxfordshire, OX29 0YD
Registered office: Westfield Road, Southam, Warwickshire CV47 0RA
www.scholastic.co.uk

Printed by Bell & Bain Ltd, Glasgow
© 2017 Scholastic Ltd
1 2 3 4 5 6 7 8 9   7 8 9 0 1 2 3 4 5 6

British Library Cataloguing-in-Publication Data
A catalogue record for this book is available from the British Library.
ISBN 978-1407-16908-8

Note from the publisher
Please use this product in conjunction with the official specification and sample assessment materials. Ask your teacher if you are unsure where to find them.

Contents

Topic 4 — GEOMETRY AND MEASURES

Topic 5 — PROBABILITY

Topic 6 — STATISTICS

How to use this book

This Revision Guide has been produced to help you revise for your 9–1 GCSE in Higher Mathematics. Broken down into topics and subtopics, it presents the information in a manageable format. Written by a subject expert to match the new specifications, it revises all the content you need to know before you sit your exams.

The best way to retain information is to take an active approach to revision. Don't just read the information you need to remember – do something with it! Transforming information from one form into another and applying your knowledge through lots of practice will ensure that it really sinks in. Throughout this book you'll find lots of features that will make your revision an active, successful process.

DOIT!

Activities that get you to turn information from one form into another so that it really embeds in your memory.

SNAPIT!

Use the Snap it! feature in the revision app to take a picture, film a video or record audio of key concepts to help them stick. Great for revision on the go!

WORKIT!

Worked examples with model solutions to help you see how to answer a tricky question.

Callouts Step-by-step guidance to build understanding.

NAILIT!

Tips written by a subject expert to help you in the revision process.

STRETCHIT!

Questions or concepts that stretch you further and challenge you with the most difficult content.

CHECKIT!

Check your knowledge at the end of a subtopic with the Check it! questions.

REVIEW IT!

Consolidate your revision with the Review it! questions at the end of every topic.

Use the All Boards Higher Mathematics Exam Practice Book alongside the Revision Guide for a complete revision and practice solution. Packed full of exam-style questions for each subtopic, along with complete practice papers, the Exam Practice Book will get you exam ready!

HOW TO REVISE!

PLAN YOUR REVISION

Get ahead by planning your revision!

Work out the **time** you have available for revising.

*Think about when you work at your best. Are you a **morning** or an **evening** person?*

Allocate **MORE TIME** for the topics you struggle with.

Revision works best in **SMALL BURSTS**, so keep sessions **SHORT AND SWEET**!

Remember to allow time to **PRACTISE** applying what you have revised.

Use your **revision app** to put together a revision timetable.

LOOK AFTER YOURSELF

Help your brain by looking after your whole body!

Take regular **breaks** from revising – your brain needs time to digest information in order to retain it.

HOTEL

Keep **hydrated** by drinking plenty of water – dehydration stops your brain from working at its full capacity.

Regular **exercise** helps stimulate the brain and will help you relax.

Get plenty of **sleep**, especially the night before an exam.

EAT WELL and limit unhealthy snacks – your brain needs fuel for memory and concentration.

Find methods of **relaxation** that work for you throughout the revision period.

BE PREPARED!

Limit potential stress on the day of an exam by getting everything you need ready the night before.

30

Highlight key pieces of written information and use different colours to classify it.

DO IT!
Take an active approach to revision. Create **revision cards**, **lists**, **mind maps**, **flow charts** and **infographics** like this one.

Annotate and summarise your class notes and revision pages.

Revise in pairs or small groups and deliver presentations on topics to each other.

SNAP IT!
Use your phone to take pictures of your revision material so that you can **revise on the go!**

REMEMBERING INFORMATION

Use different methods to digest information!

Create a safe place to **store** all of your revision notes and keep everything together.

Use **mnemonics** or songs to help learn tricky bits of information.

RETRIEVING INFORMATION

Use the information you have remembered!

REVIEW IT! Revisit the content you have already revised to confirm your understanding.

Practise applying your knowledge using exam-style questions.

READ THE QUESTION CAREFULLY!
Make sure you understand what you are being asked to do.

CHECK IT!
Make sure you know the material using simple **recall** questions.

Mark your work to help you see where you can **improve**.

FOR HIGH-MARK QUESTIONS, SPEND TIME **PLANNING** YOUR ANSWER!

Stick to the **TIME LIMITS** you will need to in the exam.

FINISH

Number

Integers, decimals and symbols

Integers

Integers are whole numbers. They include positive and negative numbers and zero. 1, 129, 0, -2, -32 are all integers.

> **WORKIT!**
>
> Which **four** numbers in this list are integers?
>
> $\frac{1}{4}$, 0, 0.125, $\frac{1}{3}$, -0.05, 120, π, $\sqrt{4}$, $\sqrt{2}$, 17
>
> The four integers are 0, 120, $\sqrt{4}$, 17.
>
> $\sqrt{4} = 2$ (or -2) and is therefore an integer.

Place value

The position of a digit in a number tells you its value.

For example, in the number 1346 the values of each digit are as follows:

Place value	1000	100	10	1
Number	1	3	4	6

So the number 1 in 1346 represents 1000, 3 represents 300, 4 represents 40 and 6 represents 6.

The values of the digits after the decimal point are:

Place value	tenth $\left(\frac{1}{10}\right)$	hundredth $\left(\frac{1}{100}\right)$	thousandth $\left(\frac{1}{1000}\right)$
Number	2	1	5

The number 0.215 represents 2 tenths, 1 hundredth and 5 thousandths. The number 0.4 represents 4 tenths. 0.4 is bigger than 0.215 because it has more tenths.

Multiplying and dividing decimals by 10, 100 and 1000

To multiply any number by 10, move the digits one place to the left. To multiply by 100, move the digits two places to the left. To multiply by 1000, move the digits three places to the left.

	1000	100	10	1		$\frac{1}{10}$	$\frac{1}{100}$	$\frac{1}{1000}$
2.358				2		3	5	8
2.358 × 10			2	3		5	8	
2.358 × 100		2	3	5		8		
2.358 × 1000	2	3	5	8				

Similarly, to divide by 10, 100 or 1000, move the digits one, two or three places to the right.

	1000	100	10	1	.	$\frac{1}{10}$	$\frac{1}{100}$	$\frac{1}{1000}$
70			7	0				
70 ÷ 10				7				
70 ÷ 100				0		7		
70 ÷ 1000				0		0	7	
1502	1	5	0	2				
1502 ÷ 10		1	5	0		2		
1502 ÷ 100			1	5		0	2	
1502 ÷ 1000				1		5	0	2

> If necessary, put zeros into the 'empty' places after the decimal point.

WORKIT!

$0.37 \times 35 = 12.95$

Without using a calculator, work out

a 37×35

$37 \times 35 = 100 \times 0.37 \times 35$

$= 100 \times 12.95 = 1295$

> 37 = 100 × 0.37, so the answer will be 100 times bigger. Move the digits in 12.95 two places to the left.

b 3.7×350

$3.7 \times 350 = 100 \times 0.37 \times 35$

$= 100 \times 12.95 = 1295$

> 3.7 = 10 × 0.37 and 350 = 10 × 35, so the answer will be 10 × 10 = 100 times bigger.

c $\frac{12.95}{350}$

$0.37 = \frac{12.95}{35}$

$\frac{12.95}{350} = \frac{12.95}{35 \times 10} = 0.37 \div 10 = 0.037$

> Divide both sides of the original calculation by 35.

> 350 = 10 × 35. Since the 350 is the denominator, the answer will be 10 times smaller.

d $\frac{1295}{37}$

$\frac{12.95}{0.37} = 35$

$\frac{1295}{37} = \frac{12.95 \times 100}{0.37 \times 100} = 35$

> 1295 = 100 × 12.95 and 37 = 100 × 0.37, and the two 100s cancel each other out.

Mathematical symbols

Mathematical symbols save time when writing. You need to remember the following symbols.

SNAPIT!   Mathematical symbols

Symbol	Meaning	Example
=	is equal to	$x + x = 2x$
≠	is not equal to	$1 + 5 \times 2 \neq 12$
>	is greater than	$-3 > -4$
<	is less than	$0 < 4$
≥	is greater than or equal to	$x \geq 4$
≤	is less than or equal to	$y \leq 6$

DOIT!

Put each symbol and its meaning on a sticky note and place them around your house to help you remember them.

NAILIT!

When learning to add/subtract directed numbers you may use number lines. At Higher level the numbers involved may be too large to use number lines so think of temperatures on a thermometer instead.

Directed numbers

Numbers can be positive or negative or zero. Numbers with a sign (i.e. positive or negative numbers) are called **directed numbers**. Numbers therefore have a size and a sign (i.e. + or −). Directed numbers can go above and below a zero value. You can see this with temperature, which can be positive, negative or zero.

Thinking of directed numbers in terms of temperature

A good way to deal with directed numbers is to think of a thermometer. A thermometer has positive values above zero and negative values below zero.

To work out −3 + 6, start at the first number (−3). The next number (+6) means that you will go up 6 units. The first 3 units of the 6 will take you to zero and the next 3 units will take you 3 units above zero to give the answer +3 or just 3.

To work out −1 − 4, start at −1 and go down 4 units (because the sign in front of the 4 is negative): −1 − 4 = −5.

> We don't usually write the sign unless it is negative.

Calculations involving more than two directed numbers

To work out the value of (i.e. evaluate) −5 + 4 − 3:
- find out the result of the first two numbers: −5 + 4 = −1
- then use the result with the final number: −1 − 3 = −4.

SNAPIT! Multiplication and division of directed numbers

+ × + = +	− × − = +
+ × − = −	− × + = −
+ ÷ + = +	− ÷ − = +
+ ÷ − = −	− ÷ + = −

Multiplication and division of directed numbers

If you multiply or divide two directed numbers that both have the same sign (i.e. both positive or both negative), the answer will be positive. If you multiply/divide two numbers with opposite signs, the answer will be negative.

✓ CHECKIT!

1 $23 \times 0.087 = 2.001$
Without using a calculator, work out
  a 23×8.7
  b 2.3×0.87
  c $\dfrac{2.001}{0.87}$
  d $\dfrac{2001}{23}$

2 $486 \times 29 = 14\,094$
Without using a calculator, work out
  a 4.86×29
  b 0.486×2.9
  c $14\,094 \div 48.6$
  d $140.94 \div 29$

3 Arrange these numbers in ascending order.

  0.12 0.012 0 -0.5 12

4 Place the correct symbol from the following list in the box. Each symbol can be used once, more than once or not at all.

  $< \quad > \quad \geq \quad \leq \quad =$

  a $\dfrac{5}{0.5} \, \square \, 10$ **b** $1\dfrac{5}{9} \, \square \, \dfrac{4}{3}$ **c** $-3 \, \square \, -1$

5 Work out
  a $-2 + 7$
  b $-3 - 5$
  c 3×-5
  d $-12 \div -2$
  e $-1 + 7 - 10$

> For help with fractions see pages 14–7.

Addition, subtraction, multiplication and division

Addition

> This is the same as writing numbers in place value tables.

To add integers:

1. Write the numbers one above the other, with the units in the same column, i.e. lined up to the right.

2. Starting with the column on the right, add up each column of numbers.
   If the total is more than 10, write the tens digit under the next column to the left.

3. Repeat for each column. If there is a digit carried from the previous column, include this in the addition.

Similarly, to add decimal numbers, line up the decimal points vertically.

WORKIT!

Find the sum of the numbers 152, 45 and 1723.

```
  1 7 2 3
    1 5 2
+     4 5
  1 9 2 0
    1 1
```

> 3 + 2 + 5 = 10. Write 0 in the first column and carry 1 into the next column.

> Make sure that you write the carry digit in the next column to the left. Remember to add it to the total of that column.

WORKIT!

Work out 97.56 + 4.054.

```
    9 7 · 5 6
+    4 · 0 5 4
  1 0 1 · 6 1 4
      1     1
```

> Line up the decimal points.

Subtraction

To subtract integers, line the numbers up as you did for addition. Again, start with the column on the right (the units column). In each column, always subtract the number on the bottom row from the number on the top row.

> The units column is a more formal name for the ones column.

WORKIT!

Subtract 129 from 647.

```
    3 1
  6 4 7
− 1 2 9
  5 1 8
```

> 10 has been added to the 7, so reduce the number in the tens column by 1. Cross out the 4 and write 3.

> 9 is larger than 7, so 'borrow' from the next column to the left. Write 1 by the 7 to make 17.

> 17 − 9 = 8

To subtract decimals, line the numbers up on the decimal point. If the number of decimal places in the two numbers is different, fill the empty places with zeros.

NAILIT!

You may have been taught to write this slightly differently. Use the method you are most comfortable with.

WORKIT!

Work out 68.03 − 5.428.

$$
\begin{array}{r}
6\overset{7}{8}\cdot\overset{1}{0}\ \overset{2}{3}\ \overset{1}{0} \\
-\ \ 5\cdot 4\ 2\ 8 \\
\hline
6\ 2\cdot 6\ 0\ 2
\end{array}
$$

Write 0 at the end of 68.03 so both numbers have 3 decimal places.

Subtract each number on the bottom row from each number on the top row. Borrow from the next column to the left when necessary.

WORKIT!

Without using a calculator, multiply 315 by 27.

$$
\begin{array}{r}
3\ 1\ 5 \\
\times\ 2\ 7 \\
\hline
2\ 2_{,}0_{3}5 \\
6\ 3_{,}0\ 0 \\
\hline
8\ 5\ 0\ 5
\end{array}
$$

315 × 7 = 2205

315 × 2 = 630, so 315 × 20 = 6300

2205 + 6300 = 8505

Multiplication

To multiply integers:

1. Write the numbers one above the other, with the units in the same column, i.e. lined up to the right.

2. Multiply the first integer by the units digit of the second integer. Write this under the line.

3. Multiply the first integer by the tens digit of the second integer. As this is the tens digit, you need to multiply the answer by 10. Write this in the next line.

4. Repeat for higher place digits in the second integer.

5. Add the results of steps 2 to 4.

To multiply decimals, for example 6.12 × 7.4:

1. Multiply the numbers without the decimal points: 612 × 74 = 45 288.

2. Count the number of decimal places in each number: 2 and 1.

3. Add these together: 3.

4. Insert the decimal point in the answer to step 1 so that the number has the number of decimal places found in step 3: 45.288.

WORKIT!

Without using a calculator, work out 7.8 × 4.1.

$$
\begin{array}{r}
7\ .\ 8 \\
\times\ \ 4\ .\ 1 \\
\hline
7\ \ 8 \\
3\ 1_{3}\ 2\ \ 0 \\
\hline
3\ 1\ .9\ \ 8
\end{array}
$$

7.8 and 4.1 each have 1 decimal place, so total number of decimal places = 2. Insert the decimal point in the answer.

Multiply the numbers as if they were 78 and 41.

NAILIT!

Check your answer by estimating: 8 × 4 = 32, so the decimal point is in the right place.

Division

WORKIT!

Without using a calculator, work out 459 ÷ 17.

$$
\begin{array}{r}
2\ 7 \\
17\overline{)4\ 5\ 9} \\
3\ 4 \\
\hline
1\ 1\ 9 \\
1\ 1\ 9 \\
\end{array}
$$

459 ÷ 17 = 27

17 doesn't divide into 4.
45 ÷ 17 = 2 remainder 11. Write 2 above the 5. Write 34 (2 × 17) below 45.

Subtract 34 from 45 and bring down the next digit (9) to make 119.

119 ÷ 17 = 7 with no remainder. Write 7 above the 9.

To divide by a decimal, multiply both numbers by 10 (or 100, 1000,…) until you are dividing by a whole number.

WORKIT!

Without using a calculator, work out 55.3 ÷ 3.5.

$$55.3 \div 3.5 = \frac{55.3}{3.5} = \frac{55.3 \times 10}{3.5 \times 10} = \frac{553}{35}$$

$$
\begin{array}{r}
1\ 5\ .\ 8 \\
35\overline{)5\ 5\ 3\ .\ 0} \\
3\ 5 \\
\hline
2\ 0\ 3 \\
1\ 7\ 5 \\
\hline
2\ 8\ 0 \\
2\ 8\ 0 \\
\end{array}
$$

Multiply both numbers by 10, so you are dividing by 35.

Insert a decimal point and zeros after the number if necessary.

DOIT!

Make a revision card for each type of calculation.

CHECKIT!

1 Without using a calculator, work out

  a 1083 + 478

  b 2445 + 89 + 513

  c 66.55 + 3.38

  d 7.08 + 4.5 + 12.343

2 Without using a calculator, work out

  a 4556 − 1737   **c** 12.935 − 4.75

  b 674 − 387   **d** 5.77 − 0.369

3 Without using a calculator, work out

  a 634 × 47   **c** 8.32 × 4.9

  b 7.7 × 3.8

4 Without using a calculator, work out

  a 1058 ÷ 23   **c** 88.5 ÷ 2.5

  b 617.4 ÷ 1.8

Using fractions

Fractions are one number divided by another number, written one over the other. Here are some terms that may be used in fraction questions. You need to remember these.

SNAP IT!   Types of fraction

Proper fractions are fractions where the top number (the numerator) is smaller than the bottom number (the denominator), for example $\frac{2}{3}$, $\frac{1}{2}$ and $\frac{15}{16}$.

Improper fractions are fractions larger than 1 where the top number (the numerator) is bigger than the bottom number (the denominator), for example $\frac{4}{3}$, $\frac{16}{5}$ and $\frac{36}{4}$.

Mixed numbers are numbers consisting of an integer and a fractional part, for example $5\frac{4}{5}$.

NAIL IT!

'Simplify' means to cancel the fraction until you can't cancel it any more.

Try to pick the largest number you can as the cancelling will be quicker. For example, 6 divides exactly into both 12 and 18, taking you directly to the answer.

STRETCH IT!

If a fraction has a 1 as the numerator, this fraction is known as the **reciprocal** of the number in the denominator. For example, $\frac{1}{7}$ is the reciprocal of 7 and $\frac{1}{x}$ is the reciprocal of x.

Cancelling fractions

Fractions should be cancelled, if possible, by dividing the numerator and denominator by the same number that divides in exactly. Keep doing this until the fraction can't be cancelled any further.

WORKIT!

Simplify the fraction $\frac{12}{18}$.

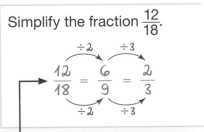

Find a number which divides exactly into the numerator and denominator. Keep going until you cannot divide by any more numbers.

Converting an improper fraction to a mixed number

To convert an improper fraction such as $\frac{5}{3}$ to a mixed number:

1. Divide the denominator (bottom number) into the numerator (top number) and write the integer down: 3 divides into 5 once with a remainder of 2, so write down 1.

2. Write the remainder over the denominator and write this next to the integer: $\frac{5}{3} = 1\frac{2}{3}$.

Converting a mixed number to an improper fraction

You often have to do this when you are multiplying or dividing fractions. To convert a mixed number such as $3\frac{2}{5}$ to an improper fraction:

1. Multiply the whole number by the denominator of the fraction: $3 \times 5 = 15$.

2. Add the numerator: $15 + 2 = 17$.

3. Put the result over the denominator: $\frac{17}{5}$.

Calculations involving fractions

Calculations involving fractions are easy if you obey some rules and are careful.

Adding fractions

You can add two fractions if they have **the same** denominator (i.e. the same number on the bottom).

- If the bottom numbers are the same, add the top numbers and then cancel if possible. For example, $\frac{5}{16} + \frac{7}{16} = \frac{12}{16} = \frac{3}{4}$.

- If the bottom numbers are **not the same**, find the common denominator (the smallest number both denominators divide exactly into).

> Both 5 and 3 divide exactly into 15. Change the first fraction to 15ths by multiplying the top and bottom by 3. Change the second fraction to 15ths by multiplying the top and bottom by 5.

WORKIT!

Work out $\frac{3}{5} + \frac{1}{3}$

$$\frac{3}{5} + \frac{1}{3} = \frac{9}{15} + \frac{5}{15} = \frac{14}{15}$$

> Once the denominators are the same, add the top numbers.

Adding two mixed numbers

To add two mixed numbers:

① Add the whole numbers.

② Add the fractions.

③ If the fractions add to an improper fraction, then convert to a mixed number.

④ Add your answer from step 1 to your answer from step 3.

WORKIT!

Work out $3\frac{3}{4} + 4\frac{3}{5}$

$$3\frac{3}{4} + 4\frac{3}{5} = 7 + \frac{3}{4} + \frac{3}{5}$$
$$= 7\frac{15}{20} + \frac{12}{20}$$
$$= 7\frac{27}{20}$$
$$= 8\frac{7}{20}$$

> $\frac{27}{20}$ is an improper fraction.
> $\frac{27}{20} = 1\frac{7}{20}$, which you add to the 7.

Subtracting fractions

To subtract fractions:

① Convert any mixed numbers to improper fractions.

② Find the common denominator of the fractions.

③ Change the fractions so the denominators are the same.

④ Subtract the top numbers of the two fractions.

⑤ Cancel if necessary.

WORKIT!

Work out $2\frac{3}{4} - 1\frac{1}{3}$

$$2\frac{3}{4} - 1\frac{1}{3} = \frac{11}{4} - \frac{4}{3}$$
$$= \frac{33}{12} - \frac{16}{12}$$
$$= \frac{17}{12}$$
$$= 1\frac{5}{12}$$

Multiplication of fractions

To multiply two fractions such as $\frac{2}{3} \times \frac{5}{8}$:

① Multiply the top numbers, then multiply the bottom numbers: $\frac{2}{3} \times \frac{5}{8} = \frac{10}{24}$.

② Check whether the fraction can be cancelled further: $\frac{10}{24} = \frac{5}{12}$.

 STRETCH IT!

You can rearrange the top numbers or the bottom numbers so that the fractions cancel and the multiplication is easier: $\frac{2}{3} \times \frac{5}{8} = \frac{2}{8} \times \frac{5}{3} = \frac{1}{4} \times \frac{5}{3} = \frac{5}{12}$.

WORKIT!

Work out

a $\frac{15}{28} \times \frac{7}{20}$

$$\frac{15}{28} \times \frac{7}{20} = \frac{15}{20} \times \frac{7}{28} = \frac{3}{4} \times \frac{1}{4} = \frac{3}{16}$$

> Rearrange the top numbers and cancel to make the multiplication easier.

b $\frac{3}{4} \times \frac{2}{15}$

$$\frac{3}{4} \times \frac{2}{15} = \frac{3}{15} \times \frac{2}{4} = \frac{1}{5} \times \frac{1}{2} = \frac{1}{10}$$

> Always check that the answer cannot be cancelled further.

c $\frac{5}{14} \times 64$

$$\frac{5}{14} \times 64 = \frac{5}{14} \times \frac{64}{1} = \frac{5}{1} \times \frac{64}{14} = \frac{5}{1} \times \frac{32}{7} = \frac{160}{7} = 22\frac{6}{7}$$

Multiplication of mixed numbers

To multiply mixed numbers such as $2\frac{1}{2} \times 1\frac{2}{5}$

1 Convert any mixed numbers to improper fractions: $2\frac{1}{2} \times 1\frac{2}{5} = \frac{5}{2} \times \frac{7}{5}$

2 Follow the steps for multiplying two fractions. Remember to simplify your answer if possible: $\frac{5}{2} \times \frac{7}{5} = \frac{35}{10} = \frac{7}{2} = 3\frac{1}{2}$

NAILIT!

Check that you can do fractions on your calculator. Most fraction questions will appear in the non-calculator paper but fractions will also appear in questions on probability. You can often use a calculator to simplify a fraction.

Division of fractions

To divide two fractions, for example $\frac{4}{5} \div \frac{2}{5}$, turn the second fraction upside down and replace the division sign with a multiplication sign.

$$\frac{4}{5} \div \frac{2}{5} = \frac{4}{5} \times \frac{5}{2} = 2$$

Division of mixed numbers

As for multiplication, first change the mixed numbers to improper fractions. Then turn the second fraction upside down and replace the division by a multiplication sign.

Ordering fractions

You may be asked to order fractions into ascending order (smallest first) or descending order (largest first). To order fractions without using a calculator, write them with a common denominator. Then they will be easy to compare.

WORKIT!

Express $3\frac{3}{4} \div \frac{5}{8}$ as a single fraction. Simplify your answer.

$$3\frac{3}{4} \div \frac{5}{8} = \frac{15}{4} \div \frac{5}{8}$$
$$= \frac{15}{4} \times \frac{8}{5}$$
$$= \frac{3}{1} \times \frac{2}{1} = 6$$

WORKIT!

Rearrange these fractions in order of size, starting from the smallest.

$$\frac{5}{6} \qquad \frac{7}{9} \qquad \frac{67}{72} \qquad \frac{2}{3}$$

$$\frac{5}{6} = \frac{60}{72} \qquad \frac{7}{9} = \frac{56}{72} \qquad \frac{67}{72} \qquad \frac{2}{3} = \frac{48}{72}$$

The order is $\frac{2}{3} \quad \frac{7}{9} \quad \frac{5}{6} \quad \frac{67}{72}$

Write all the fractions with the same denominator. The common denominator for these four fractions is 72.

NAILIT!

Make sure your answer uses the original fractions and not the ones all with the same denominator.

DOIT!

Make a set of revision cards with an example of each type of calculation.

✓ CHECKIT!

1 Here are some fractions:

$$1\frac{1}{5} \qquad \frac{16}{5} \qquad \frac{3}{4} \qquad \frac{5}{8} \qquad \frac{9}{10}$$

a Write down the fraction that is an improper fraction, and write it as a mixed number.

b Write down the fraction that is a mixed number, and write it as an improper fraction.

c Write them in ascending order.

d Add the mixed number and the improper fraction.

e Subtract the smallest number from the largest number.

2 Write down the fractions from the following list that are equivalent to the fraction $\frac{1}{3}$.

$$\frac{9}{25} \qquad \frac{15}{45} \qquad \frac{4}{12} \qquad \frac{2}{5} \qquad \frac{16}{48}$$

3 Work out the following. Give each answer as a mixed number in its simplest form.

a $3\frac{1}{3} \times 2\frac{1}{5}$ **b** $1\frac{3}{4} \div \frac{1}{2}$

4 Ravi spends $\frac{1}{3}$ of his weekly wage on rent, $\frac{1}{5}$ on bills and $\frac{1}{4}$ on food.

What fraction of his weekly wage does he have left?

Different types of number

Multiples, factors and prime factors

You need to remember these terms.

SNAPIT! Multiples and factors

- **Multiples** of a number are the answer you get when you multiply that number by another integer. For example, $3 \times 11 = 33$, so the number 33 is a multiple of 3. 33 is also a multiple of 11.

- **Factors (also called divisors)** are numbers that divide exactly into another number. For example, the factors of 24 are 1, 2, 3, 4, 6, 8, 12 and 24 as all these whole numbers divide exactly into 24. **Prime factors** are factors that are also prime numbers (see page 20).

- **Common factors** are those factors that are shared by two different numbers. The common factors of 12 and 18 are 1, 2, 3 and 6. To find common factors, list all the factors of each number and then list the factors that are in **both** lists.

- **Highest common factor (HCF)** of two numbers is the highest factor that will divide into both numbers exactly.

- **Lowest common multiple (LCM)** of two numbers is the lowest number that is a multiple of both numbers.

WORKIT!

1 List the common factors of 100 and 32.

Factors of 100:   ① ② ④ 5, 10, 20, 25, 50, 100
Factors of 32:   ① ② ④ 8, 16, 32
Common factors: 1, 2, 4

Circle the numbers that appear in both lists and write these common factors in a new list.

2 Find the highest common factor of 36 and 96.

Factors of 36:   1, 2, 3, 4, 6, 9, ⑫ 18, 36
Factors of 96:   1, 2, 3, 4, 6, 8, ⑫ 16, 24, 32, 48, 96
Highest common factor of 36 and 96 is 12.

Circle the highest number that appears in both lists.

The highest common factor is the highest number appearing in both lists.

To find the lowest common multiple, list the multiples of both numbers. For example, the multiples of 3 are 3, 6, 9, 12, 15, 18, 21, 24, 27, 30, ... and the multiples of 5 are 5, 10, 15, 20, 25, 30, ... The lowest common multiple of 3 and 5 is the lowest number in both lists: 15.

For larger numbers, you can use the following method to work out the lowest common multiple. List the prime factors for each number and then identify those factors that are common to both. Multiply the factors of the two numbers together, but only include the common factors once.

WORKIT!

a Write 28 and 36 as the products of their prime factors.

$28 = 2 \times 2 \times 7$ $36 = 2 \times 2 \times 3 \times 3$

b Find the lowest common multiple of 28 and 36.

$LCM = 2 \times 2 \times 3 \times 3 \times 7 = 252$ ◄

> Notice that the 2 × 2 part appears in both so it should be included only once in the LCM.

c Find the highest common factor of 28 and 36.

$HCF = 2 \times 2 = 4$ ◄

> The only factors that appear in both are 2 × 2.

WORKIT!

Jack is organising a barbecue. He goes to the supermarket and buys burgers in packs of 4, bread rolls in packs of 6 and sausages in packs of 8.

He buys the same number of burgers, bread rolls and sausages.

What is the minimum number of each pack he could buy? Show all your workings.

Burgers (multiples of 4)   4, 8, 12, 16, 20, ⟨24⟩ 28, 32, 36,...

Bread rolls (multiples of 6)   6, 12, 18, ⟨24⟩ 30, 36, 42, 48,...

Sausages (multiples of 8)   8, 16, ⟨24⟩ 32, 40, 48, 56,... ◄

> You need to find the multiples of each item of food and then find the lowest number that is common to all three lists (the LCM).

The lowest common multiple for the three lists is 24, giving the minimum number of packs:

6 packs of burgers   4 packs of bread rolls   3 packs of sausages

Prime numbers

A number is **prime** if it has exactly two factors: 1 and itself. Remember that 1 is not prime, because it has only one factor. This means 2 is the smallest prime number. It is the only even prime number.

Writing a number as a product of prime factors

Every integer greater than or equal to 2 can be written as a product of prime factors.

DOIT!

There are lots of terms to remember. Produce pairs of cards for each term and its definition. Practise matching the pairs of cards regularly.

Keep breaking numbers into their factors until all the resulting numbers are prime.

WORKIT!

1 Write the number 230 as the product of prime factors.

$230 = 10 \times 23 = 2 \times 5 \times 23$

2 The number 1440 can be expressed as the product of prime factors in the form $1440 = 2^p \times 3^q \times 5^r$. Find the values of the integers p, q and r.

$1440 = 144 \times 10 = 12 \times 12 \times 5 \times 2$
$= 6 \times 2 \times 6 \times 2 \times 5 \times 2$
$= 3 \times 2 \times 2 \times 3 \times 2 \times 2 \times 5 \times 2$
$= 2^5 \times 3^2 \times 5$

$p = 5, q = 2$ and $r = 1$

CHECKIT!

1 Consider the following numbers: 5, 7, 16.

   a Which number is one more than a multiple of 5?

   b Which number is one less than a factor of 12?

   c Which number is **not** a prime number?

2 Write the number 300 as the product of prime factors.

3 Three grandchildren call around to see their grandmother every 12 days, 16 days and 18 days, respectively. Find out how often all three call on the same day.

4 a When written as the product of prime factors, $360 = 2^3 \times 3^2 \times 5$.

   Write 756 as the product of prime factors.

   b Work out the highest common factor of 360 and 756.

Listing strategies

Listing strategies help solve certain types of problems such as working out combinations of items. Once the total number of combinations has been found, it is possible to work out probabilities of certain events.

WORKIT!

A restaurant offers a fixed price menu where you order any one starter and one main course from the following lists.

Starters	Main courses
Soup	Pizza
Pâté	Lasagne
	Spaghetti
Mushrooms	Salmon

Write the first item in the starters list with each main course. Repeat for the next starter. Keep doing this until all the starters have been linked.

John likes to have a different combination of starter and main course each day.
Over how many days could he have a different combination?

Soup Pizza	Soup Lasagne	Soup Spaghetti	Soup Salmon
Pâté Pizza	Pâté Lasagne	Pâté Spaghetti	Pâté Salmon
Mushrooms Pizza	Mushrooms Lasagne	Mushrooms Spaghetti	Mushrooms Salmon

This gives 12 combinations.

John could have different combinations over 12 days.

You could be asked about probabilities. For example, 'Meals are given out at random. What is the probability that John will get soup and salmon?' Soup and salmon is one of the 12 combinations worked out in the example above, so:

Probability of soup and salmon = $\frac{1}{12}$

WORKIT!

There are 5 men and 4 women in a quiz team.

a One of the men and one of the women are to be chosen as a pair to go onto a TV quiz. How many possible pairs are there?

Number of pairs = 5 × 4 = 20.

b In a different quiz two men are to be chosen. How many possible pairs of two men are there?

Call the men A, B, C, D and E.

If A was chosen first then B, C, D and E would be left for the second choice so this would give:

AB, AC, AD, AE

If B was chosen first then A, C, D and E could be possible second choices to give:

BA, BC, BD, BE

This process could be continued choosing C, D and then E as the first choice to give the following pairs.

CA, CB, CD, CE   DA, DB, DC, DE   EA, EB, EC, ED

This list reduces to:

AB, AC, AD, AE   BC, BD, BE   CD, CE   DE

This is 10 pairs.

There are 10 possible pairs of two men.

> **Each of the men could be paired with any of the 4 women.**

> **You might think that the first man can be chosen from 5 and the second man can be chosen from 4 making a total of 5 × 4 = 20 possible pairs. This would be wrong.**

> **As you are looking at pairs, the order of selection does not matter: AB and BA are the same pair.**
>
> **Strike off those that are the same pair.**

NAILIT!

If you relate the pairs to combinations of people by calling them A, B, C, etc. then this question becomes easier.

CHECKIT!

1 How many different outcomes are there when tossing a coin and rolling a six-sided dice?

2 A 3-digit number is picked.

The first number is not zero.

The 3-digit number is a multiple of 5.

How many 3-digit numbers could be picked?

The order of operations in calculations

NAILIT!

If you perform the whole calculation from left to right, you may get the wrong answer. You must remember to use BIDMAS. Do the division and multiplication before the addition and subtraction.

When following a series of calculations there is an order to be used. The word BIDMAS is used to help remember the order. BIDMAS stands for **B**rackets, **I**ndices (powers and roots), **D**ivision and **M**ultiplication, **A**ddition and **S**ubtraction. When you are evaluating (working out) a calculation, start with brackets, then indices, also known as orders (i.e. powers and roots), then division and multiplication, then addition and subtraction.

WORKIT!

Do the division and multiplication first, then the subtraction.

Work out

a $2 + 3 \times 4$ $2 + 3 \times 4 = 2 + 12 = 14$

b $5 \times 8 - 10 \div 2$ $5 \times 8 - 10 \div 2 = 40 - 5 = 35$

Do the calculation in the brackets first.

c $3 \times (24 - 9)$ $3 \times (24 - 9) = 3 \times 15 = 45$

There are no brackets so deal with indices (i.e. powers and roots) first.

d $172 - 8 \times 3^2$ $172 - 8 \times 3^2 = 172 - 8 \times 9$

$$= 172 - 72 = 100$$

e $\dfrac{21 + 7}{4}$ $\dfrac{21 + 7}{4} = \dfrac{28}{4} = 7$

The division line acts like brackets so the addition can be considered to be in brackets and should be done first.

Sometimes brackets are added for emphasis. For example these two expressions are identical.

$16 \times 3 - 24 \div 6$ $(16 \times 3) - (24 \div 6)$

WORKIT!

The calculations inside the brackets are worked out first.

The square root acts like a bracket so the calculation inside it is worked out first. Notice the sign change where two minuses are multiplied together.

Work out

a $(6 - 4)^2 + (8 \div 4)^3$ $(6 - 4)^2 + (8 \div 4)^3 = 2^2 + 2^3 = 4 + 8 = 12$

b $\sqrt{8^2 - 4 \times 3 \times (-3)}$ $\sqrt{8^2 - 4 \times 3 \times (-3)} = \sqrt{64 + 36} = \sqrt{100} = \pm 10$

DOIT!

Produce a revision poster or card for BIDMAS.

✓ CHECKIT!

1 Work out

a $10 + 4 \times 2$ **c** $(7 - 4)^2 + (8 \div 2)^2$

b $5 \times 3 - 4 \div 2$

2 Work out

a $2 + 3 \div \dfrac{1}{3} - 1$ **c** $\sqrt{4 - 3 \times (-7)}$

b $15 - (4 - 6)^3$

Indices

Indices is another name for powers. For example, in the number 2^3, the **index** is the power to which 2 is raised (3 in this case). There are rules that you need to remember when dealing with indices.

Laws of indices

The **laws of indices** are rules that apply when the same number is raised to a power. For example, the rules would apply to $2^5 \times 2^4$ because both terms in the expression are 2 raised to a power. They would not apply to $2^3 \times 5^4$ where the two numbers are different (i.e. 2 and 5).

> Note that the denominator of the fraction means take that root of the number, e.g. an index of $\frac{1}{3}$ means take the cube root.

SNAPIT! Laws of indices

Multiplication	$a^m \times a^n = a^{m+n}$
Division	$a^m \div a^n = a^{m-n}$
Powers of powers	$(a^m)^n = a^{m \times n} = a^{mn}$
Negative indices	$a^{-m} = \frac{1}{a^m}$ (provided $a \neq 0$)
Fractional indices	$a^{\frac{m}{n}} = \sqrt[n]{a^m} = (\sqrt[n]{a})^m$
Negative and fractional indices	$a^{-\frac{m}{n}} = \frac{1}{a^{\frac{m}{n}}} = \frac{1}{\sqrt[n]{a^m}}$ or $\frac{1}{(\sqrt[n]{a})^m}$
Zero indices	$a^0 = 1$ (provided $a \neq 0$)

NAILIT!

If you have an index that is a negative fraction, it is best to deal with the negative part first.

For example:

$4^{\frac{-3}{4}} = \frac{1}{4^{\frac{3}{4}}}$

$= \frac{1}{\sqrt[4]{4^3}}$

$= \frac{1}{\sqrt[4]{64}}$

$= \frac{1}{\sqrt[4]{16}\,\sqrt[4]{4}}$

$= \frac{1}{2\sqrt{2}}$

$= \frac{1}{2\sqrt{2}} \times \frac{\sqrt{2}}{\sqrt{2}}$

$= \frac{\sqrt{2}}{4}$

WORKIT!

1 Express these as single powers.

a $2^3 \times 2^5$

$2^3 \times 2^5 = 2^{(3+5)} = 2^8$

> Add 3 and 5, because $2^3 \times 2^5 = 2 \times 2 \times 2 \times 2 \times 2 \times 2 \times 2 \times 2$

b $\frac{5^5}{5^7}$

$\frac{5^5}{5^7} = 5^{(5-7)} = 5^{-2}$

> Subtract 7 from 5 because $\frac{5^5}{5^7}$
> $= \frac{5 \times 5 \times 5 \times 5 \times 5}{5 \times 5 \times 5 \times 5 \times 5 \times 5 \times 5}$
> $= \frac{1}{5^2}$

> The final index does not always have to be positive.

c $(2^{-2})^3$

$(2^{-2})^3 = 2^{(-2 \times 3)} = 2^{-6}$

NAILIT!

You need to recognise square and cube numbers.

2 Evaluate

a 2^{-3}

$2^{-3} = \frac{1}{2^3} = \frac{1}{8} = 0.125$

> You can give your answer as a fraction or a decimal, unless the question specifies.

b $27^{\frac{1}{3}}$

$27^{\frac{1}{3}} = \sqrt[3]{27} = 3$

c $8^{\frac{2}{3}}$

$$8^{\frac{2}{3}} = (\sqrt[3]{8})^2 = 2^2 = 4$$

> Alternatively you could start by writing $(\sqrt{16^3})^{-1}$

d $16^{-\frac{3}{2}}$

$$16^{-\frac{3}{2}} = \frac{1}{16^{\frac{3}{2}}} = \frac{1}{(\sqrt{16})^3} = \frac{1}{\pm 4^3} = \pm\frac{1}{64}$$

> Find the square root of 16 and then cube the answer, rather than cubing 16 first and then finding the square root.

e 5^0

$$5^0 = 1$$

> Any number to the power 0 is equal to 1.

NAILIT!

It doesn't matter whether you find the root first and then raise the answer to the power or the other way around. Generally, finding the root first will keep the numbers low and so more manageable.

Simplifying more complex expressions

You can use the rules of indices with more complex expressions: for example, expressions containing more than two terms, or expressions including powers of two different numbers.

WORKIT!

Express $\dfrac{2^{-2} \times 2^5}{2}$ as a single power of 2.

$$\frac{2^{-2} \times 2^5}{2} = \frac{2^{-2} \times 2^5}{2^1} = 2^{-2+5-1} = 2^2$$

> $2 = 2^1$

If you have an expression with powers of two different numbers, you can combine the powers of each number separately.

WORKIT!

Simplify $(3^5)^2 \times 5^2 \times 3^{-4} \times 5$.

$$(3^5)^2 \times 5^2 \times 3^{-4} \times 5 = 3^{(5 \times 2 - 4)} \times 5^{(2 + 1)}$$
$$= 3^6 \times 5^3$$

> Gather together the terms for each number and simplify.

DOIT!

Make some matching revision cards of the rules, for example $a^m \times a^n$ on one card and a^{m+n} on the matching card.

✓ CHECKIT!

1 Express each of the following as a single power.

a $7^7 \times 7^3$

b $3^{-2} \div 3^4$

c $(5^4)^5$

2 Simplify

a $\dfrac{5^4 \times 5^6}{5^3}$

b $(6^{-4} \times 6^2)^3 \div 6^{-3}$

c $5^{-7} \times 2^8 \times 5^4 \times 2^2$

d $\dfrac{7^4 \times 11^3 \times (7^{-2})^{-3}}{11^4}$

3 Find the value of

a 16^0

b $100^{\frac{1}{2}}$

c $64^{\frac{2}{3}}$

d $25^{-\frac{1}{2}}$

e $27^{\frac{1}{3}} \times 36^{\frac{1}{2}}$

4 Find the value of x such that $(3^{2x})^2 = 81$.

Surds

The number 18 is broken down into two factors, one of which is a square number. The number 9 is a perfect square, so its square root is an integer.

A surd is the square root of a non-square number. Surds are irrational numbers. This means that they cannot be expressed as fractions, recurring decimals or terminating decimals. For example, $\sqrt{5}$ is a surd.

Surds can be simplified like this:

$$\sqrt{18} = \sqrt{9 \times 2} = 3\sqrt{2}$$

STRETCH IT!

Always try to find the largest square factor. For example, $\sqrt{80}$ could be written as $\sqrt{4 \times 20}$, or as $\sqrt{8 \times 10}$ but both expressions need further simplification. It is quicker to spot that 16 is the highest square factor:

$$\sqrt{80} = \sqrt{16 \times 5} = 4\sqrt{5}$$

WORKIT!

Write $\sqrt{50}$ in the form $k\sqrt{2}$, where k is an integer.

$$\sqrt{50} = \sqrt{25 \times 2} = 5\sqrt{2}$$

Simplifying surds

Here are some general rules when manipulating surds.

SNAPIT! Simplifying surds

$$\sqrt{a} \times \sqrt{a} = a$$
$$\sqrt{a} \times \sqrt{b} = \sqrt{ab}$$
$$(\sqrt{a} + \sqrt{b})(\sqrt{a} - \sqrt{b}) = a - b$$
$$(\sqrt{a} + \sqrt{b})(\sqrt{a} + \sqrt{b}) = a + \sqrt{ab} + \sqrt{ab} + b = a + 2\sqrt{ab} + b$$
$$(a + \sqrt{b})(c + \sqrt{d}) = ac + a\sqrt{d} + c\sqrt{b} + \sqrt{bd}$$
$$a\sqrt{b} \times c\sqrt{b} = a \times \sqrt{b} \times \sqrt{b} \times c = abc$$
$$a\sqrt{b} + c\sqrt{b} = (a + c)\sqrt{b}$$

These examples show how the above rules can be applied:

1. $(\sqrt{3})^2 = \sqrt{3} \times \sqrt{3} = 3$
2. $(5\sqrt{2})^2 = 5\sqrt{2} \times 5\sqrt{2} = 25 \times 2 = 50$
3. $3\sqrt{2} \times 4\sqrt{2} = 12 \times 2 = 24$
4. $3\sqrt{2} + 2\sqrt{2} = 5\sqrt{2}$
5. $(2 + \sqrt{7})(2 + \sqrt{7}) = 4 + 2\sqrt{7} + 2\sqrt{7} + 7$
   $$= 11 + 4\sqrt{7}$$

   Use the normal rules for multiplying out pairs of brackets.

6. $(1 + \sqrt{3})(5 - \sqrt{12}) = 1(5 - \sqrt{12}) + \sqrt{3}(5 - \sqrt{12})$
   $$= 5 - 2\sqrt{3} + 5\sqrt{3} - \sqrt{36}$$
   $$= -1 + 3\sqrt{3}$$

Rationalising fractions with surds in the denominator

If you have a fraction with a surd on the bottom the surd needs to be avoided in your final answer (i.e. **rationalised**). This is done by multiplying the numerator and denominator of the fraction by the surd. Rationalising makes sure that the denominator is no longer an irrational number.

$$\frac{1}{\sqrt{3}} = \frac{1}{\sqrt{3}} \times \frac{\sqrt{3}}{\sqrt{3}} = \frac{\sqrt{3}}{3}$$

> The fraction is simplified when there are no surds in the denominator.

If the denominator has the form $a + \sqrt{b}$, for example $\frac{1}{1 - \sqrt{2}}$, remove the irrational number in the denominator by multiplying both the numerator and denominator by the expression that is the same as the denominator but with the opposite sign for the surd, in this case $1 + \sqrt{2}$.

$$\frac{1}{1 - \sqrt{2}} = \frac{1}{(1 - \sqrt{2})} \times \frac{(1 + \sqrt{2})}{(1 + \sqrt{2})} = \frac{1 + \sqrt{2}}{1 - 2} = \frac{1 + \sqrt{2}}{-1} = -1 - \sqrt{2}$$

> Rationalise the denominator by multiplying both parts of the fraction by an expression that will eliminate the surd in the denominator when the brackets are multiplied out (the **conjugate**). Usually it's the same expression with the opposite sign: here, $(2 + \sqrt{5})$.

WORKIT!

> Rationalise the denominator (i.e. elminate the surd from the denominator) and simplify the result.

1 Simplify $\frac{10}{\sqrt{5}}$.

$$\frac{10}{\sqrt{5}} = \frac{10}{\sqrt{5}} \times \frac{\sqrt{5}}{\sqrt{5}} = \frac{10\sqrt{5}}{5} = 2\sqrt{5}$$

2 Simplify $\frac{1}{2 - \sqrt{5}}$.

$$\frac{1}{(2 - \sqrt{5})} \frac{(2 + \sqrt{5})}{(2 + \sqrt{5})} = \frac{2 + \sqrt{5}}{4 - 5} = -2 - \sqrt{5}$$

3 Show that $\sqrt{3}\left(5\sqrt{12} + \sqrt{3}\right)$ can be simplified to give 33.

$$\sqrt{3}\left(5\sqrt{12} + \sqrt{3}\right) = 5\sqrt{36} + 3 = 5 \times 6 + 3 = 33$$

NAILIT!

'Show that' means show working that leads to the answer given. In part 3 (left) you are asked to show working that leads to the answer 33.

✓ CHECKIT!

1 Simplify

  a $\sqrt{3} \times \sqrt{2}$　　**c** $2\sqrt{3} \times 3\sqrt{3}$

  b $(\sqrt{5})^2$　　**d** $(2\sqrt{5})^2$

2 $\sqrt{28}$ can be written as $a\sqrt{7}$.
  Find the value of a.

3 Express as a multiple of the smallest possible surd

  a $\sqrt{45}$　　　　　**b** $\sqrt{72}$

4 Simplify

  a $\frac{16}{3\sqrt{2}}$　　**b** $\frac{18}{4 + \sqrt{7}}$

5 Simplify fully

  a $(1 + \sqrt{5})(1 - \sqrt{5})$

  b $(2 + \sqrt{3})^2$

  c $(1 + \sqrt{3})(2 + \sqrt{3})$

Standard form

Representing numbers in standard form

Standard form is used as an alternative way of writing very large or very small numbers. Numbers in standard form are represented as:

$a \times 10^n$

where $1 \leq a < 10$ (between 1 and 10, including 1 but not including 10)

n is an integer (a whole number, positive or negative or zero).

	Valid values	Invalid values
a	1, 1.2, 3.555, 9.8009	0.8, 0.25, 0.999, 10, 11.4, 200
n	1, 5, 10, 23, 0, −1, −5, −10	1.5, 0.25, −0.02, −1.5

WORKIT!

State whether each of the following numbers is in standard form.
For those that are not, explain why not.

a 1.00×10^3

Yes

b 0.75×10^{-2}

No, a is less than 1.

c 3.667×10^{-5}

Yes

d $7.5 \times 10^{2.5}$

No, n is not an integer.

e 10×10^{-100}

No, a must be less than 10.

> Look at the values of a and n in $a \times 10^n$.

Converting large numbers into standard form

To convert large integers, such as 4500, into standard form:

1 Write the digits of the number in a place value table so that it is a number between 1 and 10.

2 Work out how the digits move to make the original number: three places to the left, so multiply by 10^3.

> To multiply, move digits left.

	1000	100	10	1	$\frac{1}{10}$	$\frac{1}{100}$	$\frac{1}{1000}$
				4 • 5			
$4.5 \times 10^3 =$	4	5	0	0			

> The digits move two places to the left, so multiply by 10^2.

> Note that $16 = 1.6 \times 10^1$ but since $10^1 = 10$ this is written as 1.6×10.

WORKIT!

Convert these numbers into standard form.

a 300

 $300 = 3 \times 10^2$

b 16

$16 = 1.6 \times 10$

c 345

$345 = 3.45 \times 10^2$

d 78000

$78\,000 = 7.8 \times 10^4$

e 126.4

$126.4 = 1.264 \times 10^2$

Converting small numbers into standard form

To convert numbers smaller than 1, such as 0.0003, into standard form:

1 Write the digits of the number in a place value table so that it is a number between 1 and 10.

2 Work out how the digits move to make the original number: four places to the right, so divide by 10^4. However, in standard form you must multiply by 10^n. Dividing by 10^4 is the same as multiplying by 10^{-4}. ◄

1	.	$\frac{1}{10}$	$\frac{1}{100}$	$\frac{1}{1000}$	$\frac{1}{10\ 000}$
3					

$3 \div 10^4 = 3 \times 10^{-4} =$

| 0 | . | 0 | 0 | 0 | 3 | ◄

> This is one of the laws of indices covered on page 24.

> To divide, move digits right.

WORKIT!

Convert these numbers into standard form.

a 0.008

$0.008 = 8 \times 10^{-3}$

b 0.0000357

$0.0000357 = 3.57 \times 10^{-5}$

c 0.00000000045

$0.00000000045 = 4.5 \times 10^{-10}$

> The digits move three places to the right, so multiply by 10^{-3}.

NAILIT!

Always check that your final answer includes a number between 1 and 10 multiplied by 10 to a power.

Converting other numbers into standard form

Sometimes a number looks like it is in standard form but isn't. For example, 12×10^{-4} looks like it is in standard form because it contains a power of 10. However, the number 12 is not between 1 and 10.

To convert numbers like this into standard form:

1 Convert the number 12 into a valid value for a: $12 = 1.2 \times 10$.

2 Multiply the original power of 10 by this number:
$12 \times 10^{-4} = 1.2 \times 10 \times 10^{-4} = 1.2 \times 10^{-3}$.

NAILIT!

Check that you know how to enter standard form numbers on your calculator, and how to interpret standard form answers. The key on your calculator may be EXP, ×10ˣ or EE.

WORKIT!

Write these numbers in standard form.

a 125×10^5

$125 \times 10^5 = 1.25 \times 10^2 \times 10^5$
$\qquad\quad = 1.25 \times 10^7$

b 250×10^{-3}

$250 \times 10^{-3} = 2.5 \times 10^2 \times 10^{-3}$
$\qquad\qquad = 2.5 \times 10^{-1}$

DOIT!

Use the internet to look up some very large and very small sizes. Convert these to standard form.

Comparing numbers in standard form

To compare two standard form numbers, first compare the index parts. If these are the same, then compare the number parts. For example:

$2 \times 10^5 < 1.5 \times 10^8$

$5.83 \times 10^{-3} < 6.1 \times 10^{-3}$

Calculations using standard form

Here are some calculations that can be carried out without using a calculator.

WORKIT!

> Multiply the ordinary numbers first: 3×4. Then add the indices together to give the power of 10.

> The answer is not in standard form. Write 12 in standard form: $12 = 1.2 \times 10$ and then multiply by the power of 10.

Work out these calculations without using a calculator. Give your answers in standard form.

a $3 \times 10^5 \times 4 \times 10^2$

$3 \times 10^5 \times 4 \times 10^2 = 12 \times 10^7$

$= 1.2 \times 10 \times 10^7$

$= 1.2 \times 10^8$

b $4 \times 10^{-8} \times 4 \times 10^{-2}$

$4 \times 10^{-8} \times 4 \times 10^{-2}$

$= 16 \times 10^{-10} = 1.6 \times 10^{-9}$

c $\dfrac{8 \times 10^{-3}}{2 \times 10^{-2}}$

$\dfrac{8 \times 10^{-3}}{2 \times 10^{-2}} = 4 \times 10^{-3-(-2)}$

$= 4 \times 10^{-1}$

> 16×10^{-10} needs to be changed to standard form.

CHECKIT!

1 Write as ordinary numbers

  a 5×10^{-3} **b** 5.65×10^5

2 Express in standard form

  a $25\,000$ **c** 0.05×10^4

  b 0.00125 **d** 14×10^{-3}

3 $a = 8 \times 10^4$, $b = 3 \times 10^{-2}$ and $c = 4 \times 10^2$
Calculate the following, giving your answers in standard form.

  a b^2 **c** $\dfrac{a}{c}$

  b ab **d** $a + c$

4 a In 2016 the richest person in Britain had a wealth of £13.3 billion.

Write this amount in standard form. (1 billion is one thousand million.)

  b He decides to give his money away by giving £500 to as many people as he can. How many people could he give £500 to? Give your answer as an ordinary number.

5 a Write $15\,500$ in standard form.

  b Write 6.55×10^5 as an ordinary number.

  c Work out $\dfrac{1.25 \times 10^5}{2.5 \times 10^2}$, giving your answer in standard form.

  d Evaluate $\sqrt{1.6 \times 10^7}$, giving your answer in standard form.

Converting between fractions and decimals

Here are some familiar fractions and their decimal equivalents.

SNAP IT! Converting between fractions and decimals

$\frac{1}{2} = 0.5$ $\frac{1}{4} = 0.25$ $\frac{1}{8} = 0.125$ $\frac{1}{10} = 0.1$

$\frac{1}{20} = 0.05$ $\frac{1}{3} = 0.333...$ $\frac{2}{3} = 0.666...$

NAIL IT!

You should remember all these conversions as they often come up in the examination.

To convert a fraction, for example $\frac{4}{25}$, to a decimal, divide the bottom number (25) into the top number (4).

25 does not divide into 4. Write '0.' above the top line.

$$25\overline{)4\ .\ 0\ 0}$$
$$\begin{array}{r} 0\ .\ 1\ 6 \\ 25\overline{)4\ .\ 0\ 0} \\ 2\ 5 \\ \hline 1\ 5\ 0 \\ 1\ 5\ 0 \\ \hline 0 \end{array}$$

25 divides into 40 once, with remainder 15. Write '1' above the top line.

Bring the zero down from the number you are dividing.

NAIL IT!

Use long division in a non-calculator paper and your calculator in a calculator paper.

Terminating and recurring decimals

A **terminating** decimal has digits which do not go on forever, for example 0.25 and 1.56. In a **recurring** decimal, a number or pattern of numbers after the decimal point repeats. For example, $\frac{1}{3} = 0.333333...$, $\frac{7}{22} = 0.318181818...$ and $\frac{8}{27} = 0.296296296...$ are all recurring decimals.

To decide whether a fraction will produce a terminating or recurring decimal:

1 Cancel the fraction as far as possible.

2 Write the denominator as a product of prime factors.

3 If the prime factors in the denominator are only 2 and/or 5, then the decimal will terminate.

Converting terminating decimals to fractions

To convert a terminating decimal such as 0.625 into a fraction:

1 Count the number of digits after the decimal point.

2 Write the decimal in a place value table:

1		$\frac{1}{10}$	$\frac{1}{100}$	$\frac{1}{1000}$	$\frac{1}{10\,000}$
0	•	6	2	5	

NAIL IT!

You may be asked to put a list of numbers containing both decimals and fractions in order. First convert the fractions to decimals and then order the numbers. In your answer, give each number in the form it was given to you.

③ Use the final digit to work out the denominator: final digit (5) is thousandths, so the denominator is 1000, giving $\frac{625}{1000}$

④ Cancel the fraction as far as possible: $\frac{625}{1000} = \frac{5}{8}$

> Multiply by multiples of 10 until you have two numbers with exactly the same digits after the decimal point.

Changing recurring decimals to fractions

Recurring decimals can be written with a dot over the number if a single number repeats (e.g. $\frac{1}{3}$ 0.3333333... = 0.$\dot{3}$) or dots over the first and the last number of a block of repeating numbers (e.g. $\frac{7}{22}$ 0.318181818... = 0.3$\dot{1}\dot{8}$).

WORKIT!

Write the recurring decimal 0.8$\dot{6}$ as a fraction in its simplest form.

Let $x = 0.8\dot{6} = 0.86666666...$

$10x = 8.6666666...$

$100x = 86.666666...$

$100x - 10x = 86.666666...$
$\qquad\qquad - 8.666666...$

$90x = 78$

$x = \frac{78}{90}$

$x = \frac{13}{15}$

> Cancel the fraction by dividing both top and bottom by 6.

> Subtract $10x$ from $100x$ to eliminate the digits after the decimal point.

WORKIT!

Write the recurring decimal 0.$\dot{3}\dot{9}$ as a fraction in its simplest form.

Let $x = 0.\dot{3}\dot{9} = 0.39393939...$

$100x = 39.39393939...$

$100x - x = 39.39393939...$
$\qquad\qquad - 0.39393939...$

$99x = 39$

$x = \frac{39}{99} = \frac{13}{33}$

> Subtract x from $100x$ to eliminate the digits after the decimal point.

DOIT!

> Make cards of the numbers 1–20. Shuffle the cards and pick two to make a fraction then work out whether the fraction is a recurring or terminating decimal.

NAILIT!

If one of these questions appears in the calculator paper, divide the fraction to check you get back to the original recurring decimal.

CHECKIT!

1 Express these fractions as decimals.

   a $\frac{43}{100}$ b $\frac{3}{8}$ c $\frac{11}{20}$

2 Express these terminating decimals as fractions in their simplest form.

   a 0.8   b 0.45   c 0.584

3 Express these recurring decimals as fractions in their simplest form.

   a 0.$\dot{7}$ b 0.0$\dot{4}$ c 0.9$\dot{5}\dot{4}$

4 Express these decimals as fractions in their simplest form.

   a 0.$\dot{5}1\dot{8}$ b 0.76

5 Show that 0.$\dot{7}$ + $\frac{2}{9}$ = 1.

6 Put these numbers in order, starting with the smallest.

   0.85, $\frac{7}{8}$, $\frac{4}{5}$, −0.9, $\frac{7}{10}$

Converting between fractions and percentages

A **percentage** is a fraction where the denominator is 100. For example, $\frac{6}{100} = 6\%$.

Percentages allow values to be compared, as they are always out of 100.

Converting a percentage to a fraction

To convert a percentage to a fraction, divide the percentage by 100 and then simplify the fraction if possible.

WORKIT!

Convert 45% to a fraction in its simplest form.

$45\% = \frac{45}{100} = \frac{9}{20}$

Converting a fraction to a percentage

If possible, multiply the top and bottom of the fraction so that the bottom number is 100. The top number is then the percentage.

For example, $\frac{2}{5} = \frac{2 \times 20}{5 \times 20} = \frac{40}{100} = 40\%$.

Alternatively, convert the fraction to a decimal and then multiply the decimal by 100.

For example, $\frac{1}{6} = 0.16666... = 16.67\%$.

WORKIT!

Convert $\frac{36}{55}$ to a percentage. Give your answer to two decimal places.

$\frac{36}{55} \times 100 = 65.45\%$ ◄———

You can also multiply the fraction by 100 and then work out the answer.

NAILIT!

Failing to cancel fractions will lose you marks. Always look at the fraction carefully to see if it can be cancelled.

NAILIT!

This method works when the denominator is a factor of 100.

DOIT!

Create a table of conversions between common fractions and percentages.

✓ CHECKIT!

1 Convert these percentages to fractions.

   a 25%   **b** 85%   **c** 68%

2 Charlie scored $\frac{65}{80}$ in a maths test and 80% in an English test. Which test did he do better in? Show your working.

3 Convert these fractions to percentages. Where necessary, give your answer to 1 decimal place.

   a $\frac{3}{10}$ **b** $\frac{4}{25}$ **c** $\frac{3}{7}$

Fractions and percentages as operators

NAILIT!

These types of questions could come up on a calculator or a non-calculator paper. Use a calculator if it is a calculator paper, as it is usually not possible to cancel the fractions easily.

Fractions and percentages are often used as **multipliers** when solving problems.

> Replace 'of' with a multiplication sign.

WORKIT!

Work out

a 22% of £600

$$22\% \text{ of } £600 = \frac{22}{100} \times 600 = 22 \times 6 = £132$$

b $\frac{21}{20}$ of £600

$$\frac{21}{20} \text{ of } £600 = \frac{21}{20} \times 600 = 21 \times 30 = £630$$

NAILIT!

This question would be asked in a non-calculator paper so you need to be able to cancel the fraction.

WORKIT!

In school A, 48% of the 600 students are boys.

In school B, $\frac{4}{9}$ of the 630 students are boys.

Which school has the greater number of boys and by how many?

$$48\% \text{ of } 600 = \frac{48}{100} \times 600 = 48 \times 6 = 288$$

$$\frac{4}{9} \text{ of } 630 = \frac{4}{9} \times 630 = 4 \times 70 = 280$$

School A has the greater number of boys by 8.

> Find 48% of 600 and $\frac{4}{9}$ of 630. Remember that 'of' means multiply.

NAILIT!

Always specifically state the answer. The working out is not enough for you to gain full marks.

WORKIT!

Joshua spends $\frac{1}{3}$ of his weekly wage on rent, $\frac{1}{4}$ on food and $\frac{1}{5}$ on entertainment. How much money does he have left each week if he earns £480 per week?

Rent: $\frac{1}{3} \times 480 = £160$ Food: $\frac{1}{4} \times 480 = £120$

Entertainment: $\frac{1}{5} \times 480 = £96$

Amount remaining = $480 - (160 + 120 + 96) = £104$

CHECKIT!

1 a Find $\frac{3}{4}$ of £640.

   b Find 15% of £30.

   c Find 195% of 80 kg.

2 In School A 56% of the pupils are boys and in School B 35% of the pupils are girls. There are 600 pupils in School A and 700 pupils in School B.

How many boys are there in each school?

Standard measurement units

Many quantities can be expressed in different units. For example, time can be expressed in seconds, minutes, hours, days or years and distance can be expressed in metres or kilometres. It is very important to be able to convert quantities to different units.

Metric unit conversions

SNAPIT! Metric unit conversions

1 cm = 10 mm	1 tonne = 1000 kg
1 m = 100 cm	1 cm³ = 1 ml
1 km = 1000 m	1 litre = 1000 cm³
1 kg = 1000 g	1 litre = 1000 ml

WORKIT!

Convert

a 3.3 litres into ml

3.3 litres = 3.3 × 1000 = 3300 ml ◄─── To convert from a larger unit to a smaller unit, always multiply.

b 850 g into kg

850 g = 850 ÷ 1000 = 0.85 kg ◄─── To convert from a smaller unit to a larger unit, always divide.

c 2 m into mm.

2 m = 2 × 100 = 200 cm

2 m = 200 cm = 200 × 10 = 2000 mm ◄─── First convert m to cm by multiplying by 100. Then multiply by 10 to convert cm to mm.

NAILIT!

You can also convert straight from m to mm, by multiplying by 1000.

Time and money

SNAPIT! Time

60 seconds = 1 minute	7 days = 1 week
60 minutes = 1 hour	52 weeks = 1 year
24 hours = 1 day	

Money is usually either in pounds or pence.

WORKIT!

1   A large tin of beans costs 67p.

Tins of beans are packed into trays of 24 tins. There are 20 trays on a pallet and 40 pallets are loaded onto a van.

Find the value of all the beans loaded onto the van, giving your answer in pounds.

One tin of beans costs £0.67.

Cost of beans on van = 0.67 × 24 × 20 × 40 = £12 864

2   Ahmed earns a salary of £28 000 per year for a 40-hour week.

a   Work out his monthly wage.

$$\text{Monthly wage} = \frac{\text{Yearly wage}}{12} = \frac{28\,000}{12} = £2333.33$$

b   Calculate his daily rate for the month of June. Give your answer correct to the nearest penny.

$$\text{Daily rate for June} = \frac{\text{Monthly wage}}{30} = \frac{2333.33}{30} = £77.78$$

> **Convert the 67p to pounds by dividing by 100.**

> Assume he gets the same amount each month.

NAILIT!

Never write two units next to a number, for example £2.35p. It will be marked incorrect, as it is not clear whether it is £2.35 or 2.35p.

✓ CHECKIT!

1   Convert

a   9.7 kg to grams

b   850 cm^3 to litres

c   2.05 km to centimetres

2   Work out how many seconds there are in one day. Write your answer in standard form.

3   A café sells a carton of juice for 34p.

It has 12 boxes, each containing 20 cartons.

Work out the value of the café's stock of juice. Give your answer in pounds.

Rounding numbers

Rounding numbers

Rather than give exact values of numbers, we often approximate them. In calculations you will be given values that need to be put into an equation to calculate an unknown value. You may be asked to give the answers to a certain number of **decimal places** or **significant figures**.

Decimal places

123.764 is given to 3 decimal places (d.p.) as there are three digits to the right of the decimal point.

To give a number to a specified number of decimal places, look at the digit in the next decimal place. For example, to round to 2 decimal places, look at the digit in the 3rd place after the decimal point:

- If this is less than 5, then the digit in the 2nd place stays the same.
- If it is 5 or greater, the digit in the 2nd place increases by 1.

The 5th digit after the decimal point is 8, so round the 4th place up to 8.

This is written as 1.0 rather than just 1 to show that the number is given to one decimal place.

WORKIT!

1   Write 4.56378 to

   a 4 decimal places   b 2 decimal places

      4.5638   4.56

2   Write 1.03598 to

   a 1 decimal place   b 4 decimal places

      1.0   1.0360

The 3rd digit after the decimal point is 3, so the 2nd place digit stays the same.

The 5th digit after the decimal place is 8, so the 9 in the 4th place needs to increase. So the 9 becomes 0 and the next digit to the left increases by 1.

Significant figures

Another way of rounding a number is to express it as significant figures (s.f.). This means a specified number of consecutive digits, starting with a non-zero digit.

SNAPIT! Significant figures

Here are some values and the number of significant figures they are written to:

273	3 significant figures	0.0002	1 significant figure
4200	2 significant figures	0.076	2 significant figures
8.31	3 significant figures	0.1005	4 significant figures

The counting for the significant figures starts at the first non-zero digit.

Zeros at the beginning and end of numbers are not usually significant: 4200 is to 2 s.f. and 0.0002 is to 1 s.f. However, in the middle of numbers zeros are significant as they contribute to the size of the number: 0.1005 is to 4 s.f.

WORKIT!

Give these numbers to the stated number of significant figures.

Count the numbers, starting with the first non-zero digit. Once you reach the required significant figure, use the next digit to round up or down.

Fill any positions to the right of the last significant digit with zeros, so the number still has the same number of digits.

a 3.5609   2 s.f.   3.6

b 94236   3 s.f.   94 200

c 0.0040658   3 s.f.   0.00407

Starting counting from the first non-zero digit, the first 3 significant figures are 406.

NAILIT!

Always read the question to see whether the answer should be given to a certain number of significant figures or decimal places or as a whole number.

DOIT!

Draw a flowchart for rounding to a certain number of decimal places.

CHECKIT!

1 Write each number to 3 significant figures.

a 1259

b 14.919

c 0.0003079

d 9 084 097

e 1.8099×10^{-4}

2 Write each number to the number of decimal places specified in brackets.

a 10.565 (1)

b 123.9765 (3)

c 0.02557 (2)

d 3.9707 (3)

e 0.00195 (3)

f 4.098 (2)

3 Here is a number, 1989.

a Write this number to 1 significant figure.

b Write this number to 2 significant figures.

c Write this number to 3 significant figures.

4 Write the number 3.755×10^{-4} as an ordinary number to 4 decimal places.

Estimation

Estimating answers

It is easy to make a mistake when entering numbers into a calculator, so it is important to have an idea what the answer should be. This means that you can recognise a wrong answer. Estimating is an important part of checking your answers.

When making estimations, write all the values in the calculation to 1 significant figure and give the answer to 1 significant figure.

WORKIT!

Estimate the answers to these calculations, giving your answers to 1 significant figure. Compare your estimates with the exact answers using your calculator.

a $2.8 \times 0.9 \times 3.4$

$2.8 \times 0.9 \times 3.4 \approx 3 \times 1 \times 3 = 9$ ◄

> Write all the numbers to 1 significant figure. The symbol $\approx$ means approximately equal to.

Working this out accurately on a calculator, $2.8 \times 0.9 \times 3.4$
$= 8.568$ which is approximately 9.

b $\dfrac{2.68 \times 7.89 \times 12.00}{2.98 \times 6.07 \times 11.67}$

$\dfrac{2.68 \times 7.89 \times 12.00}{2.98 \times 6.07 \times 11.67} \approx \dfrac{3 \times 8 \times 10}{3 \times 6 \times 10} = \dfrac{4}{3} = 1.\dot{3} = 1$ (to 1 s.f.) ◄

> Cancel the numbers in the numerator with those in the denominator as far as possible.

Working this out accurately on a calculator, $\dfrac{2.68 \times 7.89 \times 12.00}{2.98 \times 6.07 \times 11.67}$

$= 1.2020$ which is approximately 1.

c $\sqrt{2.5 \times 4.1 \times 2.9}$

$\sqrt{2.5 \times 4.1 \times 2.9} \approx \sqrt{3 \times 4 \times 3} = \sqrt{36} = 6$ or -6 (1 s.f.)

Working this out accurately on a calculator, $\sqrt{2.5 \times 4.1 \times 2.9}$

$= \pm 5.4521$ which is approximately 5 or -5.

Estimating square roots and powers

To estimate the value of $\sqrt{32}$ to 1 decimal place, start by finding perfect squares that are above and below the number inside the square root: $5^2 = 25$ and $6^2 = 36$, so the answer is between 5 and 6. Then decide whether 32 is nearer to 36 or 25. It's nearer to 36, so the answer is probably nearer to 6. A reasonable estimate might be 5.7. ◄

> $\sqrt{32}$ is 5.656854249, so the estimate was pretty good.

Powers can be estimated in a similar way. To estimate $(3.6)^3$, work out the whole number powers on either side: $3^3 = 27$ and $4^3 = 64$. The value of $(3.6)^3$ ◄ will be a little over midway between them, so estimate 45.

> $(3.6)^3$ is 46.656.

WORKIT!

Estimate values for

a $\sqrt{90}$ to 1 decimal place

$9^2 = 81$ and $10^2 = 100$ so $\sqrt{90}$ lies between 9 and 10.

90 lies almost midway between 81 and 100, so 9.5 is a reasonable estimate.

$\sqrt{90} \approx 9.5$ (to 1 d.p.)

> The actual value is 9.487 (to 3 d.p.).

b $(5.25)^2$ to the nearest whole number

$5^2 = 25$ and $6^2 = 36$

5.25 is closer to 5, so estimate 28

$(5.25)^2 \approx 28$ (to nearest whole number)

> The actual value is 27.5625.

c $5^{2.8}$ to the nearest whole number.

$5^2 = 25$ and $5^3 = 125$

2.8 is closer to 3, so estimate

$5^{2.8} \approx 95$ (to nearest whole number)

> The actual value is 90.597 (to 3 d.p.)

> Allow for the fact that as the power gets larger, the values are further apart. So $5^{2.5}$ is closer to 5^2 than to 5^3.

✓ CHECKIT!

1 Work out an estimate for

a $\dfrac{5.9 \times 189}{0.54}$

b $\sqrt{4.65} + 28.9 \div 6$

2 Use estimation to find which of the answers A, B, or C is closest to the exact answer. Approximate each number in the calculation to one significant figure.

	Calculation	Answer A	Answer B	Answer C
a	$3.45 \times 2.78 \times 0.09$	0.8	1	0.08
b	$12.56 \times 1.87 \times 0.45$	6	26	13
c	$120 \div 0.45$	300	200	60
d	$0.01 \times 0.15 \times 109$	0.3	0.2	2
e	$0.12 \times 300 \times 0.53$	15	41	9
f	$6.07 \times 3.67 \times 0.1$	4	2	5
g	$20.75 \div 6.98$	0.3	4	3
h	$0.01 \times 145 \times 35$	40	400	450
i	$6.5 \times 0.3 \times 0.01$	0.01	0.02	0.2
j	$65 \div 1050$	0.07	0.1	0.001

3 Estimate values for

a $\sqrt{45}$ to 1 decimal place

b $\sqrt{104}$ to 1 decimal place

c $(2.3)^3$

d $3^{1.4}$

Upper and lower bounds

Finding upper and lower bounds

If you measured the length of a line to the nearest mm using a ruler, the measurement could be up to 0.5 mm larger or smaller. The maximum value a measurement can have is called the **upper bound** and the minimum value is called the **lower bound**.

For a measured length of 45 mm:

upper bound = 45 + 0.5 = 45.5 mm

lower bound = 45 − 0.5 = 44.5 mm

WORKIT!

The mass of coffee in a packet is 227 g to the nearest g. Find the upper and lower bounds.

Upper bound = 227 + 0.5 = 227.5g

Lower bound = 227 − 0.5 = 226.5g

> Find half of the given unit (i.e. 1 g). Add it to the value for the upper bound and subtract it for the lower bound.

Error intervals

The **error interval** is the range of values (between the upper and lower bounds) in which the precise value could lie. It is normally written as:

lower bound ≤ x < upper bound

Truncation occurs when all the digits past a certain point are cut off.

For example, the fraction $\frac{2}{3}$ = 0.66666... truncated to 2 decimal places becomes 0.66.

WORKIT!

A length, l cm, is recorded as 2.54 cm.

a Write the error interval for l where the measurement has been truncated to 3 significant figures.

Upper bound = 2.549999... cm

> 2.549999... is equivalent to 2.55.

Lower bound = 2.54 cm

The error interval is 2.54 ≤ l < 2.55 cm

b Write the error interval for l where the measurement has been rounded to 2 decimal places.

$\frac{0.01}{2}$ = 0.005 cm

> This is half the smallest unit in the given measurement.

Lower bound = 2.54 − 0.005 = 2.535 cm

Upper bound = 2.54 + 0.005 = 2.545 cm

The error interval is 2.535 ≤ l < 2.545 cm

DOIT!

Draw a poster to show how to find the upper and lower bounds of the volume of a cuboid.

Calculations using upper and lower bounds

WORKIT!

$T = \dfrac{\sqrt{l}}{g}$ $l = 2.658$ correct to 4 significant figures

$g = 9.81$ correct to 3 significant figures

a   By considering the upper bounds and the lower bounds for l and g, work out the error interval for T.

Give a reason for your answer.

l has an upper bound of 2.6585 and a lower bound of 2.6575

g has an upper bound of 9.815 and a lower bound of 9.805

Maximum value of $T = \dfrac{\sqrt{l}}{g} = \dfrac{\sqrt{2.6585}}{9.805} = 0.16629$ ◄——

> The maximum value of T will be when the upper bound is used for l and the lower bound for g.

Minimum value of $T = \dfrac{\sqrt{l}}{g} = \dfrac{\sqrt{2.6575}}{9.815} = 0.16609$ ◄——

> The minimum value of T will be when the lower bound is used for l and the upper bound for g.

$0.16609 \leq T < 0.16629$

b   Give a value for T to a suitable degree of accuracy.

The upper and lower bounds for T are the same when written to 3 significant figures.

So $T = 0.166$ to 3 significant figures

> To find the upper and lower bounds of values calculated from two numbers, you need to use the bounds shown in the table.

> The smallest value of $A + B$ will be made from the smallest values of both A and B.

SNAPIT!   Upper and lower bounds

Operation	Calculation	Lower bound of calculation		Upper bound of calculation	
		A	B	A	B
Addition	$A + B$	Lower	Lower	Upper	Upper
Subtraction	$A - B$	Lower	Upper	Upper	Lower
Multiplication	$A \times B$	Lower	Lower	Upper	Upper
Division	$A \div B$	Lower	Upper	Upper	Lower

✓ CHECKIT!

1   The length of a piece of wood is 145 cm to the nearest centimetre.

   a   Write down the lower bound for the length of the piece of wood.

   b   Write down the upper bound for the length of the piece of wood.

   c   Express these bounds as an error interval for the length of wood, l.

1 **a** Using a calculator, evaluate $5\sqrt{2.43 \times 10^7}$.

   b Using a calculator, evaluate $10^{\frac{4}{3}}$. Give your answer to 1 decimal place.

2 Without using a calculator, work out

   a $8 - 3 \div 2$

   b $(-1)^3 + 32 \div 8 + 1$

   c $\dfrac{27}{\frac{1}{3}}$ **d** 1.65×3.6

3 Without using a calculator, simplify the following fractions, giving your answers as mixed numbers in their simplest form.

   a $\dfrac{64}{12}$ **b** $\dfrac{124}{13}$

4 Write the numbers 2, 3 and 4 in the boxes to give the largest possible total. Each number can only be used once.

$$\square \dfrac{\square}{4} + \dfrac{1}{\square}$$

5 A rectangle has a length of 2.23 cm and a width of 4.55 cm, both to the nearest 0.01 cm.

   a Work out the lower and upper bounds for the area of the rectangle in cm² to 3 decimal places.

   b Give the area of the rectangle to a suitable degree of accuracy.

6 In 18 g of water there are 6.02×10^{23} molecules.

   a How many molecules would there be in 10 g of water? Give your answer in standard form to 3 significant figures.

   b Find the mass in kg of one molecule of water. Give your answer in standard form to 3 significant figures.

7 Work out an estimate for $(0.45 \times 0.78)^2$.

8 Find the value (as an integer or fraction) of

   a 7^0 **b** $9^{\frac{1}{2}}$ **c** 8^{-2} **d** $64^{-\frac{1}{3}}$

9 Dylan rounds a number y to one decimal place. On rounding the number, the result is 5.6. Write down the error interval for y.

10 Show that the expression $\dfrac{1 - \sqrt{2}}{1 + \sqrt{2}}$ simplifies to $2\sqrt{2} - 3$.

11 The width of a sheet of glass, a, is 112 cm correct to 3 significant figures. Write down the error interval for a.

12 Write the recurring decimal $0.\dot{7}\dot{2}$ as a fraction in its simplest form.

13 $c = \dfrac{a}{b}$

   $a = 0.6754$ correct to 4 significant figures

   $b = 2.34$ correct to 3 significant figures

   a Work out the upper and lower bounds for a and b and hence work out the error interval for c. Give a reason for your answer.

   b Give a value for c to a suitable degree of accuracy.

14 Work out each of the following. Simplify your answer where possible.

   a $\dfrac{3}{25} \div \dfrac{9}{50}$ **b** $25 \div \dfrac{5}{16}$

15 Write each of the following numbers in standard form:

   a 0.00000045   **c** 5640

   b 12 million

16 Without using a calculator, work out the value of $(8 \times 10^{-5}) \times (4 \times 10^3)$ giving your answer in standard form.

17 **a** Write down all the factors of 64.

   b Find the highest common factor of 64 and 100.

Algebra

Simple algebraic techniques

There are conventions (rules that everyone follows) when using algebra.

SNAP IT! Algebraic conventions

Write:

- ab instead of $a \times b$
- $3a$ instead of $a + a + a$ and $3 \times a$
- b^2 instead of $b \times b$
- a^3 instead of $a \times a \times a$
- a^2b instead of $a \times a \times b$
- $\frac{a}{b}$ instead of $a \div b$
- $\frac{1}{2}x$ or $\frac{x}{2}$ instead of $x \div 2$.

Words used in maths

When using algebra certain words are used. Here are the ones you need to know about.

SNAP IT! Algebra terms

Equations contain letters and/or numbers and they always contain an equals sign. Examples include $2x + 4 = 2$, $3x - 4 = x + 6$, $x^2 + 2x + 1 = 0$.

Expressions contain letters and/or numbers but there is no equals sign. Examples include $2x - y^2$, $(x + 3)(x - 2)$.

Identities are true for all values of the letter or letters.

Examples include $(x + 1)^2 \equiv x^2 + 2x + 1$ or $(x - 2)^2 \equiv x^2 - 4x + 4$.

Formulae contain letters and/or words to represent variable quantities (**variables**) and they also contain an equals sign.

Terms are a single number or variable, or numbers and variables multiplied together. Terms are always separated by $+$ or $-$ signs. In the expression $5ab - 2a + 4$, $5ab$, $-2a$ and 4 are all terms.

Factors are what you can multiply together to get an expression in algebra. For example, $2x(x + 3) = 2x^2 + 6x$ so $2x$ and $(x + 3)$ are both factors of $2x^2 + 6x$.

The symbol $\equiv$ means 'is equivalent to'.

Notice that the left-hand side is identical to the right-hand side, so this could not be solved like an equation.

Collecting like terms

Like terms are terms that have identical letters and powers. To simplify an expression you can collect like terms (find the sum of them). For example, $4x^2y$ can be combined with $-14x^2y$, but it cannot be combined with $-2xy^2$ or $5xy$.

WORKIT!

Simplify these expressions.

> Rearrange with the like terms grouped together.

> Always give answers in descending powers (e.g. x^3, then x^2, then x and finally ordinary numbers (e.g. 2)).

a $4x^2 + 6x - x + 6x^2 + 3$

$4x^2 + 6x - x + 6x^2 + 3 = 4x^2 + 6x^2 + 6x - x + 3$

$= 10x^2 + 5x + 3$

b $6a - b - c + a - 4a + c$

$6a - b - c + a - 4a + c = 6a + a - 4a - b - c + c = 3a - b$

> Always look at both the sign and the value of a letter or number.

c $4yx + x^2 + 6xy - 4x^2$

$4yx + x^2 + 6xy - 4x^2 = 4xy + 6xy + x^2 - 4x^2 = 10xy - 3x^2$

> Although the order of the letters in a term does not matter ($4yx$ is the same as $4xy$), the convention is to write the number and then the letters in alphabetical order.

Substituting numerical values into formulae and expressions

This is the sort of calculation you often do in other subjects such as physics and chemistry. Replace each letter in the formula or expression with the numerical value given in the question and calculate.

WORKIT!

In the formula $p = \frac{nRT}{v}$, find the value of p if $n = 5$, $R = 8.31$, $v = 3$ and $T = 298$. Give your answer to 3 significant figures.

$p = \frac{nRT}{v} = \frac{5 \times 8.31 \times 298}{3}$

$= 4127.3$

$= 4130$ (to 3 s.f.)

> Write the equation and replace each letter by its value.

> Always check the question to see how many decimal places or significant figures should be used.

DOIT!

Write an example of an equation, expression, etc. on different cards. Write the terms on another set of cards. Practise matching the cards.

WORKIT!

$s = ut + \frac{1}{2}at^2$ Find the value of s when $u = 25$, $t = 2.5$ and $a = -10$.

$s = 25 \times 2.5 + \frac{1}{2} \times (-10) \times 2.5^2$

$= 62.5 - 31.25$

$= 31.25$

> Notice the use of BIDMAS.

NAILIT!

$s = ut + \frac{1}{2}at^2$ is called an equation of motion, where s is distance, u is initial velocity, t is time and a is acceleration.

CHECKIT!

1 Identify whether each of the following is a formula, expression, equation or identity.

a $s = ut + \frac{1}{2}at^2$
d $(x^2y)^3 = x^6y^3$
b $2(x^2 + y^2) = 2x^2 + 2y^2$ **e** $2x + 1 = 3$
c $4x^3y^2$

2 Simplify

a $15x^2 - 4x + x^2 + 9x - x - 6x^2$

b $7a + 5b - b - 4a - 5b$
c $8yx + 5x^2 + 2xy - 8x^2$
d $\frac{1}{2}x^3 + 3x - 5 + 2\frac{1}{2}x^3 - 4x$

3 If $P = I^2R$, find P when $I = \frac{2}{3}$ and $R = 36$.

4 Using $v = u + at$, find v when $u = 20$, $a = -8$ and $t = 2$.

Removing brackets

Multiplying out a single bracket

Multiply each term inside the bracket by the term outside the brackets. Be careful when there is a negative outside the bracket as it will change the signs of all the terms inside the bracket.

WORKIT!

Multiply out the brackets and simplify your answer.

a $3(2x + 4)$

> Multiply each term in the bracket by the number outside the bracket.

$3 \times 2x + 3 \times 4 = 6x + 12$

b $-6(x - 1)$

$-6 \times x + (-6) \times (-1) = -6x + 6$

> The minus sign changes the sign of every term inside the bracket.

c $5(2x - 3) + 4(x - 1)$

$5 \times 2x + 5 \times (-3) + 4 \times x + 4 \times (-1)$
$= 10x - 15 + 4x - 4$
$= 14x - 19$

> Multiply out the brackets, then collect like terms.

Using the laws of indices

You have already used the laws of indices with numbers (see page 24). The same laws apply to variables.

NAILIT!

You may need to use laws of indices when multiplying out brackets.

WORKIT!

Expand and simplify

a $2x(3x + 7)$

$2x \times 3x + 2x \times 7 = 6x^2 + 14x$

> Remember that $x = x^1$.

b $4a(2a^2 - 3a + 4)$

$4a \times 2a^2 + 4a \times (-3a) + 4a \times 4 = 8a^3 - 12a^2 + 16a$

c $5x^2y^3z(3x + 7yz)$

$5x^2y^3z \times 3x + 5x^2y^3z \times 7yz = 15x^3y^3z + 35x^2y^4z^2$

WORKIT!

Simplify:

a $4m^{\frac{3}{2}} \times 2m^{-3}$

$4m^{\frac{3}{2}} \times 2m^{-3} = 4 \times 2 \times m^{\frac{3}{2} + (-3)} = 8m^{-\frac{3}{2}}$

> Multiply the two numbers and add the indices.

b $\dfrac{10c^{-\frac{4}{3}}}{2c^{-4}}$

$\dfrac{10c^{-\frac{4}{3}}}{2c^{-4}} = (10 \div 2) \times c^{-\frac{4}{3} - (-4)} = 5c^{\frac{8}{3}}$

> Divide the two numbers and subtract the indices.

c $5x^2y^3z\left(3x^{\frac{1}{2}} + 7y^{-\frac{1}{2}}z\right)$

$5x^2y^3z\left(3x^{\frac{1}{2}} + 7y^{-\frac{1}{2}}z\right) = 5x^2y^3z \times 3x^{\frac{1}{2}} + 5x^2y^3z \times 7y^{-\frac{1}{2}}z$

$= 15x^{\frac{5}{2}}y^3z + 35x^2y^{\frac{5}{2}}z^2$

> Multiply each term inside the bracket by the term outside the brackets.

Multiplying brackets

When multiplying pairs of brackets, all the terms in the first bracket must be multiplied by all the terms in the second bracket. There are different methods you can use to multiply two pairs of brackets together.

Method 1: the 'face' method

Using the 'face' method, the curved lines show that all the terms are multiplied together.

$(x + 2)\ (x - 3)$

Start from the top first and work from left to right:

- multiply x by x to give x^2

- multiply $+2$ by -3 to give -6

- multiply x by -3 to give $-3x$

> Now go to the bottom curved lines.

- multiply $+2$ by x to give $+2x$.

So, $(x + 2)(x - 3) = x^2 - 6 - 3x + 2x$
$= x^2 - x - 6.$

> Notice that the answer is given in descending powers of x.

Method 2

Each term in the first bracket multiplies the contents of the second bracket:

$(x + 2)(x - 3) = x(x - 3) + 2(x - 3)$

$= x^2 - 3x + 2x - 6$

> Multiply out each bracket.

$= x^2 - x - 6$

> Simplify by collecting terms in x.

We will use Method 2 in this book, but you should use the method you are happier with.

> The curved lines look like a face (eyebrows, nose and mouth) and remind you what to multiply.

NAILIT!

Try both methods and use the one you find easier.

NAILIT!

With practice, you may find that you do not need to write the first step, but can go straight to multiplying out each bracket.

WORKIT!

Expand

> 'Expand' means 'multiply out'.

a $(x + 5)(x - 2)$

$x(x - 2) + 5(x - 2) = x^2 - 2x + 5x - 10$

$= x^2 + 3x - 10$

> Collect like terms in x.

b $(2x - 7)(4x - 2)$

$2x(4x - 2) - 7(4x - 2) = 8x^2 - 4x - 28x + 14$

$= 8x^2 - 32x + 14$

> Remember to keep the sign with the second term from the first bracket. You need to multiply each term in the second bracket by -7.

c $(a + b)^2$

$a(a + b) + b(a + b) = a^2 + ab + ab + b^2$

$= a^2 + 2ab + b^2$

$(a + b)^2$ means $(a + b)(a + b)$.

To expand an expression containing three sets of brackets, expand the first pair of brackets. Then expand this expression with the third bracket.

WORKIT!

Expand $(2x - 1)(x + 5)(3x - 2)$.

It is acceptable to use 1s in your working out.

$[2x(x + 5) - 1(x + 5)](3x - 2)$

$= (2x^2 + 10x - x - 5)(3x - 2)$

First expand $(2x - 1)(x + 5)$.

$= (2x^2 + 9x - 5)(3x - 2)$

$= 2x^2(3x - 2) + 9x(3x - 2) - 5(3x - 2)$

Multiply each term in $2x^2 + 9x - 5$ by each term in $3x - 2$.

$= 6x^3 - 4x^2 + 27x^2 - 18x - 15x + 10$

$= 6x^3 + 23x^2 - 33x + 10$

DOIT!

Write out the expansion of two brackets and make notes or write instructions beside each line of your working.

CHECKIT!

1 Multiply out

  a $2(x + 4)$ **d** $x(3x - 1)$

  b $7(9x + 3)$ **e** $3x(x + 1)$

  c $-(1 - x)$ **f** $4x(5x - 2)$

2 Multiply out and simplify

  a $2(x + 3) + 3(x + 2)$

  b $6(x + 4) - 3(x - 7)$

  c $x(3x + 1) + x(x + 1)$

  d $x(3x - 4) - 2(3x - 4)$

3 Simplify:

  a $p^{-\frac{4}{3}} \times p^{\frac{8}{3}}$

  b $\dfrac{r^5 \times r^{-2}}{r^4}$

  c $\left(b^{\frac{3}{2}}\right)^{-4}$

4 Expand and simplify

  a $(t + 3)(t + 5)$

  b $(x - 3)(x + 3)$

  c $(2y + 9)(3y + 7)$

  d $(2x - 1)^2$

5 Expand and simplify

  a $(x + 7)(x + 2)(2x + 3)$

  b $(2x - 1)(3x - 2)(4x - 3)$

Factorising

Factorisation is the opposite process to expanding brackets.

Factorising a simple expression

Any factors that are common to all terms in an expression can be taken outside a bracket. These may be letters or numbers. When the terms include numbers (**coefficients**) the highest common factor of these is taken outside.

The highest common factor of $9p$ and $15q$ is 3.

x and y are common to both terms, so write xy outside the bracket.

WORKIT!

1   Factorise $9p + 15q$.

$3(3p + 5q)$

'Factorise fully' means that **all** the common factors need to be taken outside the bracket.

2   Factorise fully

a $x^2y - xy$

$xy(x - 1)$

b $15a^2b - 12ab$

$3(5a^2b - 4ab) = 3ab(5a - 4)$

Factorising a quadratic expression

Factorising a quadratic expression gives two linear expressions in brackets. For example, $x^2 - 3x + 2$ factorises to $(x - 1)(x - 2)$.

Linear expressions contain the variable to the power 1 and no higher: for example, $2x - 7$.

There are many different methods for factorising a quadratic expression, one of which is shown below. This method is particularly useful when the numbers in the quadratic expression are large.

Suppose you are asked to factorise $x^2 + 7x - 30$.

You need to find two factors that multiply together to give -30. They also need to add together to give the coefficient of x ($+7$ in this case).

The coefficient of x is the number in front of x.

State the factor pairs of -30:

1 and -30	3 and -10	6 and -5
-1 and 30	-3 and 10	-6 and 5

Then look for the pair of numbers in the above list that add together to give 7:

-3 and 10.

With practice you may not need to write the list out.

Now write the two factors you have found as the coefficients of the x terms like this:

$x^2 + 7x - 30 = x^2 - 3x + 10x - 30$

Then factorise the first two terms by taking a common factor of x out and also the last two terms by taking a number out like this:

> **The terms inside the brackets have to be identical.**

$x^2 - 3x + 10x - 30 = x(x - 3) + 10(x - 3)$

The term $(x - 3)$ is common, so may be taken out as a common factor in the following way:

> **If you had written the factors the other way round it would still work:**
> $x^2 + 10x - 3x - 30$
> $= x(x + 10) - 3(x + 10)$
> $= (x + 10)(x - 3).$

$x(x - 3) + 10(x - 3) = (x - 3)(x + 10)$

This may seem long-winded, but it works with all quadratic expressions that can be factorised, as the following examples show.

WORKIT!

Factorise $x^2 - 9x + 18$.

$x^2 - 9x + 18 = x^2 - 6x - 3x + 18$

$= x(x - 6) - 3(x - 6)$

$= (x - 6)(x - 3)$

> **The factors of 18 that add together to give -9 are -6 and -3. These are written as the coefficients of the two x-terms.**

> **The coefficient of x^2 is the number in front of x^2.**

If the coefficient of x^2 is not 1, multiply it by the number term. Then look for factor pairs of this number.

WORKIT!

Factorise $6x^2 - 17x + 12$.

$6x^2 - 17x + 12 = 6x^2 - 9x - 8x + 12$

$= 3x(2x - 3) - 4(2x - 3)$

$= (2x - 3)(3x - 4)$

> **Notice that the sign for the number outside the bracket is negative. This makes the contents of both brackets identical.**

> **$6 \times 12 = 72$**
> **Look for a factor pair of 72 that adds to give -17: -9 and -8.**

The difference of two squares

A perfect square is a number or expression whose square root does not contain a square root symbol. For example, 16, $36a^2$ and $100x^2y^4$ are all perfect squares; 10, $16b^3$ and $9xy^2$ are not perfect squares.

If two terms are both perfect squares with a minus sign between them, such as $x^2 - y^2$, this is called the difference of two squares. They can be factorised like this: $x^2 - y^2 = (x + y)(x - y)$.

Both terms must be perfect squares, so capable of being easily square rooted. Note that this only works when there is a minus sign between the two terms. For example:

$4x^2 - 9y^2 = (2x + 3y)(2x - 3y)$

$16x^2 - 25 = (4x + 5)(4x - 5)$

$4x^2 - 10 = (4x + \sqrt{10})(4x - \sqrt{10})$

NAILIT!

Always check your factorisation by multiplying out the brackets. It is easy to make a mistake, especially with signs.

> **Square root each term and put them in brackets: one bracket has a + sign between the terms and the other has a − sign.**

SNAPIT!

Difference of two squares

Learn the formula for the difference of two squares:

$a^2 - b^2 \equiv (a + b)(a - b)$

Factorising to simplify

Sometimes the question does not state that you need to factorise, but it is necessary in order to simplify an expression.

WORKIT!

1 Show that $\frac{1}{9x^2 - 1} \div \frac{1}{3x^2 + 20x - 7}$ simplifies to $\frac{ax + b}{cx + d}$ where a, b, c and d are integers.

$$\frac{1}{9x^2 - 1} \div \frac{1}{3x^2 + 20x - 7} = \frac{3x^2 + 20x - 7}{9x^2 - 1}$$

> Express the division as a single fraction.

> Factorise both the numerator and the denominator.

$$= \frac{(3x - 1)(x + 7)}{(3x + 1)(3x - 1)}$$

$$= \frac{x + 7}{3x + 1}$$

> Cancel the common factor.

2 Write $\frac{1}{x + 1} + \frac{x}{x - 1}$ as a single fraction.

$$\frac{1}{x + 1} + \frac{x}{x - 1} = \frac{1(x - 1) + x(x + 1)}{(x + 1)(x - 1)}$$

> Find the common denominator, which will be the denominators multiplied together.

$$= \frac{x - 1 + x^2 + x}{(x + 1)(x - 1)}$$

$$= \frac{x^2 + 2x - 1}{(x + 1)(x - 1)}$$

> The numerator will not factorise, so there are no common factors to cancel.

DOIT!

Write a quick summary of how to factorise algebraic expressions.

CHECKIT!

1 Factorise fully

a $24t + 18$

b $9a - 2ab$

c $5xy + 15yz$

d $24x^3y^2 + 6xy^2$

2 Factorise

a $x^2 + 10x + 21$

b $x^2 + 2x - 15$

c $6x^2 + 19x + 10$

d $4x^2 - 49$

e $25x^2 - 7$

3 Simplify

$$\frac{2x^2 + 7x - 4}{x^2 + 7x + 12}$$

4 Work out

$$\frac{1}{x - 7} - \frac{x + 10}{2x^2 - 11x - 21}$$

Express your answer in its simplest form.

Changing the subject of a formula

Changing the subject of a formula helps you to find a particular quantity. You need to get the letter representing that quantity on the left of the equals sign and everything else on the right. For example:

$$n = \frac{m}{M}$$

As n is on its own on the left-hand side, it is the **subject** of the formula. To change the subject of the formula to m, you need to remove M from the right-hand side. Multiply both sides by M to give:

$$Mn = \frac{Mm}{M}$$

Cancel the M in the top and bottom of the fraction on the right-hand side:

$$Mn = m$$

Swap this around to give:

$$m = Mn$$

To make M the subject of $n = \frac{m}{M}$, first multiply both sides by M to give:

$$Mn = m$$

Then remove the n from the left-hand side by dividing both sides by n.

So, $M = \frac{m}{n}$

The upper case and lower case letters mean different things, so they are separate terms.

We normally write the letters in an equation in alphabetical order: Mn rather than nM.

m is now the subject.

M is now the subject.

WORKIT!

Rearrange the following equations to make the bracketed letter the subject of the formula.

a $y = \frac{1}{x}$ (x)

$xy = 1$

$x = \frac{1}{y}$

b $y = 2x + 6$ (x)

$y - 6 = 2x$

$x = \frac{y - 6}{2}$

c $r^2 = a^2 + b^2$ (b)

$r^2 - a^2 = b^2$

$b^2 = r^2 - a^2$

$b = \sqrt{r^2 - a^2}$

Multiply both sides by x to remove x from the denominator.

Divide both sides by y.

Subtract 6 from both sides.

Divide both sides by 2.

Subtract a^2 from both sides.

Square root both sides.

SNAPIT!

Rearranging a formula for a term that appears on both sides

When the new subject appears on both sides of a formula:

1. If there are any brackets, multiply them out.

2. Get all the terms that contain the subject on one side of the equation and all the other terms on the other side.

3. Take the new subject out as a common factor so you will have it multiplied by the contents of a bracket.

4. Divide both sides by the bracket, leaving the subject on its own on the left of the equation.

WORKIT!

Make y the subject of the formula $3(3x - y) = 9 - 4xy$.

Collect the terms containing the new subject on one side and all the other terms on the other side.

Multiply out the brackets.

$$9x - 3y = 9 - 4xy$$

$$4xy - 3y = 9 - 9x$$

$$y(4x - 3) = 9 - 9x$$

Take the new subject, y, out of a bracket as a factor.

Divide both sides by the contents of the bracket to leave the new subject on its own.

$$y = \frac{9 - 9x}{4x - 3}$$

WORKIT!

Make b the subject of the following formula.

$$\sqrt{8a + 4b} = 4a$$

$$8a + 4b = 16a^2$$

Square both sides to remove the square root.

$$4b = 16a^2 - 8a$$

$$b = \frac{16a^2 - 8a}{4}$$

$$b = 4a^2 - 2a$$

$$b = 2a(2a - 1)$$

WORKIT!

Here is a formula used in physics. Rearrange it to make g the subject.

$$T = 2\pi\sqrt{\frac{l}{g}}$$

$$T^2 = \frac{4\pi^2 l}{g}$$

Square both sides to remove the square root.

$$gT^2 = 4\pi^2 l$$

$$g = \frac{4\pi^2 l}{T^2}$$

WORKIT!

Express x in terms of y for the following equation.

$$\frac{x}{2} - 7y = -6x$$

$$x - 14y = -12x$$

$$13x = 14y$$

$$x = \frac{14y}{13}$$

Multiply both sides by 2 to remove the 2 in the denominator.

To change the subject of a formula that includes algebraic fractions, first multiply through by the denominators. For example, to make y the subject of

$$x = \frac{1}{y} + \frac{2}{z + 1}:$$

$$xy(z + 1) = (z + 1) + 2y$$

Then rearrange as described above:

$$xy(z + 1) - 2y = z + 1$$

$$y(xz + x - 2) = z + 1$$

to give $y = \dfrac{z + 1}{xz + x - 2}$

DO IT!

Draw a flowchart for rearranging a formula when the subject appears on both sides.

CHECKIT!

1 Make r the subject of the formula.

  a $A = \pi r^2$

  b $A = 4\pi r^2$

  c $V = \frac{4}{3}\pi r^3$

2 Make the bracketed symbol the subject of the formula.

  a $y = mx + c$ (c) **d** $v^2 = u^2 + 2as$ (u)

  b $v = u + at$ (a) **e** $s = \frac{1}{2}(u + v)t$ (t)

  c $v^2 = 2as$ (s) **f** $\frac{1}{x} = \frac{1}{2p} + \frac{1}{q}$ (x)

Solving linear equations

NAIL IT!

Remember:
$x \equiv x^1$

Linear equations are ones where the unknown or **variable** doesn't have a power (except 1). The variable is usually called x, but it can be any letter or symbol. You only need one linear equation to solve for one variable.

Remember that an equation is like a balance. If you apply a process to one side (such as add a number), then for the equation to remain in balance, the same process (adding the same number) must be applied to the other side.

WORKIT!

Solve these equations.

a $x + 5 = 7$

$x + 5 - 5 = 7 - 5$

$x = 2$

> You want to get x on its own on the left-hand side.
> Subtract 5 from both sides.

b $4x = 24$

$x = 24 \div 4 = 6$

> To remove the 4 in front of the x, divide both sides by 4.

c $4x - 7 = 21$

$4x = 28$

$x = 7$

> First, get a term in x on its own. Add 7 to both sides.

d $\dfrac{2x - 7}{3} = 7$

$2x - 7 = 21$

$2x = 28$

$x = 14$

> Remove the denominator by multiplying both sides by 3.

Linear equations involving brackets

When solving linear equations containing brackets, first multiply out the brackets and collect any like terms. Then solve the equation in the way outlined in the previous examples.

WORKIT!

Solve these equations.

a $3(2x - 3) = 0$

$6x - 9 = 0$

$6x = 9$

$x = 1.5$

> Multiply out the brackets.

b $5(x - 3) - 3(x + 1) = 0$

$5x - 15 - 3x - 3 = 0$

$2x - 18 = 0$

$2x = 18$

$x = 9$

> Multiply out the brackets and collect like terms.

NAIL IT!

It is better to rearrange the equation in such a way that the unknown quantity is positive.

Unknown quantity on both sides of the equation

If the unknown quantity (i.e. the quantity you are asked to find) appears on both sides of the equation, you must get it on just one side.

WORKIT!

Solve these equations.

a $2x - 1 = x + 4$

$x - 1 = 4$

$x = 5$

> You could subtract $2x$ from both sides but this would give $-x$ on the right-hand side. Subtracting x from both sides is a better option.

b $3(3x - 5) = 12(x - 7)$

$9x - 15 = 12x - 84$

$-15 = 3x - 84$

$69 = 3x$

$23 = x$

$x = 23$

> Subtract $9x$ from both sides.

> Subtract 84 from both sides.

WORKIT!

Solve these equations.

a $1 - 2x = 4(x - 2)$

$1 - 2x = 4x - 8$

$1 = 6x - 8$

$9 = 6x$

$x = \dfrac{9}{6} = \dfrac{3}{2} = 1.5$

b $\dfrac{(2a - 1)}{2} = \dfrac{(1 - 3a)}{3}$

$3(2a - 1) = 2(1 - 3a)$

$6a - 3 = 2 - 6a$

$12a - 3 = 2$

$12a = 5$

$a = \dfrac{5}{12}$

> Multiply both sides by 6 (to remove the denominator).

NAILIT!

Solving linear equations is very important in GCSE mathematics as it happens in different kinds of questions. It is essential that you master all the techniques here as they will be needed elsewhere.

DOIT!

Make up some linear equations. Solve them tomorrow. Are your answers all integers?

CHECKIT!

1 Solve these equations.

a $x - 7 = -4$

b $9x = 27$

c $\dfrac{x}{5} = 4$

2 Solve these equations.

a $3x + 1 = 16$

b $\dfrac{2x}{3} = 12$

c $\dfrac{3x}{5} + 4 = 16$

3 Solve these equations.

a $5(1 - x) = 15$

b $-2m - 4 = -3m - 3$

c $9(4x - 3) = 3(2x + 3)$

Solving quadratic equations using factorisation

A quadratic equation is an equation that can be expressed in the form:

$ax^2 + bx + c = 0$

where a, b and c are numbers.

To solve a quadratic using factorisation, first put the equation into the form above and then factorise it. For example, to solve $2x^2 = 5 - 9x$, first rearrange it as $2x^2 + 9x - 5 = 0$ and then factorise it to give:

$(2x - 1)(x + 5) = 0$

For this to equal zero, one of the brackets must equal zero:

Hence $2x - 1 = 0$ or $x + 5 = 0$,

giving $x = \dfrac{1}{2}$ or $x = -5$

> **Before factorising the quadratic expression must be equal to zero.**

NAIL IT!

You need to be able to solve simple linear equations to solve quadratic equations by factorisation.

WORKIT!

Solve $2x^2 + 7x - 4 = 0$

$(2x - 1)(x + 4) = 0$ ← Factorise $2x^2 + 7x - 4$.

$2x - 1 = 0$ or $x + 4 = 0$ ← Put each bracket equal to 0.

$x = \dfrac{1}{2}$ or $x = -4$

WORKIT!

This rectangle has an area of 60. Find the value of x.

Area of rectangle = length × width

$= 6x(2x + 1)$

Substitute area = 60.

$60 = 6x(2x + 1)$

$60 = 12x^2 + 6x$

$0 = 12x^2 + 6x - 60$

$12x^2 + 6x - 60 = 0$ ← Divide both sides by 6 to make the factorisation easier.

$2x^2 + x - 10 = 0$

$(2x + 5)(x - 2) = 0$

$x = -\dfrac{5}{2}$ or $x = 2$ ← Put each bracket equal to 0 and solve.

$x = -\dfrac{5}{2}$ would give a length of $6 \times \left(-\dfrac{5}{2}\right) = -15$, which is impossible.

So $x = 2$

(rectangle diagram: top side labelled $6x$, right side labelled $2x + 1$)

NAIL IT!

Always check to see if both values are possible or only one of them. In some situations, you can ignore a negative value, for example if it is a length.

WORKIT!

The rectangle shown below has length $(x + 8)$ cm and width $(x - 2)$ cm. The area of the rectangle is $56\,cm^2$. Find the value of x.

Area $= (x + 8)(x - 2) = x^2 + 6x - 16$

Also, area $= 56$ so $x^2 + 6x - 16 = 56$

Hence $x^2 + 6x - 72 = 0$

Factorising we obtain $(x - 6)(x + 12) = 0$ so $x = 6$ or -12

$x = -12$ would give a negative length and width so this is impossible.

So $x = 6\,cm$

[rectangle diagram: top side labelled $x + 8$, right side labelled $x - 2$]

DOIT!

Draw a poster showing the solution of a quadratic equation using factorisation, annotating the stages.

CHECKIT!

1 Solve

a $x^2 + 5x + 6 = 0$

b $x - x^2 = -12$

c $2x^2 + 17x + 35 = 0$

2 A right-angled triangle has base $(2x + 3)$ cm and height $(x + 4)$ cm. The area of the triangle is $9\,cm^2$.

a Use the area of the triangle to show that $2x^2 - 11x - 6 = 0$.

b Solve this equation to find the only possible value of x.

c Work out the base and height of the triangle.

3 Three sides of a right-angled triangle are 13 cm, $(x + 1)$ cm and $(x + 8)$ cm. The longest side is the 13 cm side.

Calculate the value of x.

Solving quadratic equations using the formula

NAILIT!

This formula will not be given so you will need to remember it.

Not all quadratic equations can be factorised easily. There are three other methods you can use: using a formula, completing the square or drawing a graph. Completing the square is covered on page 79 and finding solutions graphically on pages 80–1.

NAILIT!

Check your answers by substituting the values back into the equation.

 SNAPIT! **Quadratic formula**

Quadratic equations of the form $ax^2 + bx + c = 0$ can be solved using the formula:

$$x = \frac{-b \pm \sqrt{b^2 - 4ac}}{2a}$$

> Be careful with signs when you are entering numbers into this formula.

NAILIT!

If you are asked to leave your answer in surd form, then you must use either the formula or the method involving completing the square (see page 79).

WORKIT!

Solve the equation $2x^2 - 5x + 1 = 0$ leaving your answers in surd form.

$x = \dfrac{-b \pm \sqrt{b^2 - 4ac}}{2a}$ > Write down the formula.

$x = \dfrac{-(-5) \pm \sqrt{(-5)^2 - 4(2)(1)}}{2(2)}$

$= \dfrac{5 \pm \sqrt{25 - 8}}{4} = \dfrac{5 \pm \sqrt{17}}{4}$

So $x = \dfrac{5 + \sqrt{17}}{4}$ or $\dfrac{5 - \sqrt{17}}{4}$

> Comparing the equation with $ax^2 + bx + c = 0$ gives $a = 2$, $b = -5$ and $c = 1$.
>
> Substitute these values into the formula.

WORKIT!

a Show that $\dfrac{3}{x + 7} = \dfrac{2 - x}{x + 1}$ can be written as $x^2 + 8x - 11 = 0$.

$\dfrac{3}{x + 7} = \dfrac{2 - x}{x + 1}$ > Multiply both sides by $(x + 1)(x + 7)$.

$3(x + 1) = (2 - x)(x + 7)$

$3x + 3 = 2x + 14 - x^2 - 7x$ > Expand the brackets.

$3x + 3 = -x^2 - 5x + 14$

$x^2 + 8x - 11 = 0$ > Rearrange in the form $ax^2 + bx + c = 0$.

b Hence solve the equation

$\frac{3}{x+7} = \frac{2-x}{x+1}$, giving your answers to 2 decimal places.

$$x = \frac{-b \pm \sqrt{b^2 - 4ac}}{2a}$$

$$= \frac{-8 \pm \sqrt{8^2 - 4(1)(-11)}}{2(1)}$$

Substitute $a = 1$, $b = 8$ and $c = -11$ into the formula.

$$= \frac{-8 \pm \sqrt{64 + 44}}{2}$$

$$= \frac{-8 \pm \sqrt{108}}{2}$$

$$= \frac{-8 + \sqrt{108}}{2} \quad \text{or} \quad \frac{-8 - \sqrt{108}}{2}$$

$$= 1.1962 \quad \text{or} \quad -9.1962$$

$$= 1.20 \text{ or } -9.20 \text{ (to 2 d.p.)}$$

NAILIT!

If the question asks for the answer to be given to a certain number of decimal places, you need to use the formula rather than factorisation.

DOIT!

Use the quadratic formula to solve the equations you solved by factorisation in the previous section.

CHECKIT!

1 Solve the equation $2x^2 - x - 7 = 0$, giving your answers correct to 3 significant figures.

2 a Show that $\frac{2x + 3}{x + 2} = 3x + 1$ can be written as $3x^2 + 5x - 1 = 0$.

b Hence solve the equation

$\frac{2x + 3}{x + 2} = 3x + 1$, giving your answers to 2 decimal places.

Solving simultaneous equations

Simultaneous equations are sets of equations that have more than one variable (e.g. x and y). Solving the equations means finding values for x and y that are true for both equations.

SNAPIT! Simultaneous equations

There are three methods that can be used to solve a pair of simultaneous equations:

1. **By elimination**: eliminate one of the unknowns by adding or subtracting the two simultaneous equations.

2. **By substitution**: substitute the expression for x or y from one of the equations into the other equation.

3. **Graphically**: use both equations to plot two lines – their point of intersection is the solution.

Solving simultaneous equations by elimination

Eliminate one of the variables by either adding or subtracting the equations.

WORKIT!

Solve the simultaneous equations.

$$2x + 3y = 8 \qquad (1)$$

$$x + 4y = 9 \qquad (2)$$

Number the equations so it is easier to refer to them.

Make the coefficients of either x or y the same for both equations, by multiplying one or both of the equations by numbers.

In this case, multiply equation (2) by 2 to make the coefficients of x the same. Label your new equation '(3)'.

$$2x + 8y = 18 \qquad (3)$$

Subtract (1) from (3).

$$5y = 10$$

$$y = 2$$

$$2x + 3y = 8 \qquad (1)$$

$$2x + 3(2) = 8$$

Substitute $y = 2$ into equation (1).

$$2x + 6 = 8$$

$$2x = 2$$

$$x = 1$$

$$x + 4y = 9 \qquad (2)$$

$$1 + 4(2) = 9$$

Check the answer by substituting $x = 1$ and $y = 2$ into equation (2).

Solution is $x = 1$ and $y = 2$.

NAILIT!

Always check your answer by substituting into the equations.

Solving simultaneous equations by substitution

This method of solving simultaneous equations involves removing y by substituting the y value from one of the equations into the other equation. The substitution method is preferred when finding the solutions to a non-linear equation and a linear equation.

> Non linear means it has a power of x, as in a quadratic equation.

WORKIT!

Solve the simultaneous equations.

$x^2 + y^2 = 16$

> The first equation represents a circle because it is the form $x^2 + y^2 = r^2$. The second equation is a straight line because it is in the form $ax + by = c$.

$x + y = 4$

$x = 4 - y$

> Rearrange the second equation to express x in terms of y.

$(4 - y)^2 + y^2 = 16$

> Now substitute for x into the first equation.

$16 - 8y + y^2 + y^2 = 16$

$2y^2 - 8y = 0$

> Subtract 16 from both sides.

$2y(y - 4) = 0$

> Remove $2y$ as a factor.

$2y = 0$ giving $y = 0$, or $y - 4 = 0$ giving $y = 4$.

Substituting $y = 0$ into $x + y = 4$ gives $x = 4$.

> You can use either equation but this one is easier.

Substituting $y = 4$ into $x + y = 4$ gives $x = 0$.

So the solutions are $x = 4$ and $y = 0$, or $x = 0$ and $y = 4$.

NAILIT!

Make sure you write 'or' between the two sets of values, not 'and'.

WORKIT!

Solve the simultaneous equations

$y = \dfrac{6}{x}$ $x + y = 7$

$\qquad x + y = 7 \qquad (1)$

$\qquad\quad y = \dfrac{6}{x} \qquad (2)$

> Number the equations so you can refer to them.

From equation (1) we have $y = 7 - x$

Substituting this value of y into equation (2) we have

$\qquad 7 - x = \dfrac{6}{x}$

Multiplying both sides by x gives

$\qquad 7x - x^2 = 6$

Rearranging to form a quadratic gives

$\qquad x^2 - 7x + 6 = 0$

$\qquad (x - 6)(x - 1) = 0$

$\qquad\qquad x = 1 \text{ or } 6$

Substituting each of these x values into equation (1) we obtain

When $x = 1$, $y = 7 - 1 = 6$

When $x = 6$, $y = 7 - 6 = 1$

Solutions are $x = 1$, $y = 6$ and $x = 6$, $y = 1$

Solving simultaneous equations graphically

Any pair of equations can be solved graphically by drawing them accurately and finding any point or points of intersection.

WORKIT!

Solve these simultaneous equations graphically, giving your solutions to 1 decimal place.

$y = x^2 - 4x + 1$

$y = x - 2$

> When asked to plot a graph, create a table of values.

x	0	1	2	3	4	5
$y = x^2 - 4x + 1$	1	-2	-3	-2	1	6

> This is a quadratic equation, so you need five or six pairs of coordinates to plot the graph.

x	0	2	5
$y = x - 2$	-2	0	3

> You only need two pairs of coordinates for a straight line, but a third one provides a check.

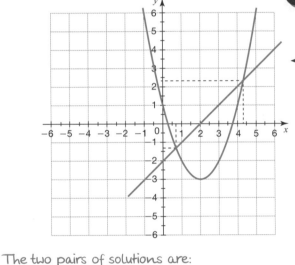

> Plot the graphs of the two equations and find where they intersect.

The two pairs of solutions are:

$x = 4.3, y = 2.3$ and $x = 0.7, y = -1.3$

STRETCHIT!

Not all simultaneous equations can be solved. If lines are parallel to each other there are no points of intersection.

DOIT!

Create revision cards to show when each method of solving simultaneous equations is most appropriate.

NAILIT!

Two linear equations could be solved in a similar way, but there would be only one point of intersection so only one pair of values.

CHECKIT!

1 Solve these pairs of simultaneous equations.

   a $y = 3x - 7$

   $y = 3 - 2x$

   b $y - 2x = -6$

   $y - 14 = -3x$

2 Solve the simultaneous equations.

$y = 10x^2 - 5x - 2$

$y = 2x - 3$

3 The line $y = 6x + 2$ and the curve $y = x^2 + 5x - 4$ intersect at two points. Use an algebraic method to find the coordinates of these points.

Solving inequalities

Representing inequalities

Number lines and set notation

Both number lines and set notation can be used to represent inequalities, as shown below.

> In set notation, { } contains the possible set of values.
>
> ∪ means that the set of numbers contains all the numbers represented by both inequalities.

$x < 25$ $\{x : x < 25\}$

5   10   15   20   25   30   35   40

> The blue line shows x is less than 25. The open circle at 25 shows that x cannot be equal to 25.

$x \geq 10$ $\{x : x \geq 10\}$

2   4   6   8   10   12   14   16   18   20

> The blue line shows x is 10 or greater. The closed circle at 10 shows that x can be equal to 10.

$-2 \leq x < 3$ $\{x : -2 \leq x < 3\}$

−4 −3 −2 −1   0   1   2   3   4   5

$x < 1$ $x \geq 6$ $\{x : x < 1 \cup < x \geq 6\}$

−1   0   1   2   3   4   5   6   7   8   9

> When there is a gap between the circles, two separate inequalities are represented.

Graphically

Inequalities can be shown graphically using boundary lines and shading. If the inequality allows points on the boundary line, then it is drawn as a solid line; otherwise it is a dotted line.

WORKIT!

Shade the region of points on a graph where

a $x \leq 3$

> Draw the boundary line $x = 3$, i.e. the line where x always has a value of 3.

> Identify the region where x always has a value less than 3. Shade this region.

NAILIT!

Always remember that the region that is allowed is the shaded region.

b $y > 5$

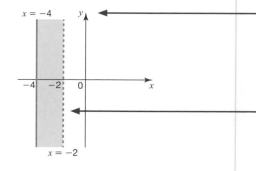

The boundary line $y = 5$ is dotted, as the inequality is $>$.

c $-4 \leq x < -2$

> Draw both boundary lines. Remember that $x = -4$ needs to be solid and $x = -2$ dotted.

> The shaded region is where x always has a value between -4 and -2.

Linear inequalities

Solve inequalities in the same way as you solve equations, by doing the same operation to both sides of the inequality.

WORKIT!

Solve these inequalities.

a $4x - 5 > 2$

> Add 5 to both sides.

$$4x > 7$$

> Divide both sides by 4.

$$x > \frac{7}{4}$$

c $-3 \leq 2x + 1 < 7$

$$-4 \leq 2x < 6$$

> Subtract 1 from all three elements.

> Divide all three elements by 2.

$$-2 \leq x < 3$$

b $3(2x - 7) \geq 12x + 15$

$$6x - 21 \geq 12x + 15$$

$$-6x - 21 \geq 15$$

$$-6x \geq 36$$

$$x \leq -6$$

> When you multiply or divide by a negative number, reverse the inequality.

Linear inequalities with two variables

Sometimes you will be given two or three equations with two variables and asked to show the solution on a graph:

> To draw a straight line, substitute two values of x into the equation to find y, and draw the line through the two points.

1. Draw the lines representing boundary lines by replacing the inequality symbol with $=$. If the inequality is $\leq$ or $\geq$ make the line solid; if it is $<$ or $>$ make the line dotted.

2. For each line, shade the region that is allowed.

> If you are not sure which side of the line to shade, substitute a pair of values for x and y.

WORKIT!

a Illustrate the region represented by the following inequalities on a graph by shading the region that is required.

$$y \leq x \qquad\qquad x + 3y \leq 12 \qquad\qquad y > 1$$

b List the integer values of x and y that satisfy all of these inequalities.

> Draw the lines $y = x$, $x + 3y = 12$ as solid lines and $y = 1$ as a dotted line.

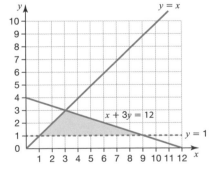

> To work out which side of $x + 3y = 12$ to shade, substitute $x = 1$ and $y = 4$: $1 + 3(4) = 13$. The inequality is $x + 3y \leq 12$, so $(1, 4)$ is not allowed.

> Any point in the required region or on solid lines enclosing the region satisfies all three inequalities.

$$(2, 2), (3, 2), (4, 2), (5, 2), (6, 2), (3, 3)$$

Quadratic inequalities

To solve a quadratic inequality such as $x^2 + x \geq 6$:

1 Ensure that the quadratic inequality has all the terms on one side and a zero on the other side: $x^2 + x - 6 \geq 0$.

2 Consider the case where $x^2 + x - 6 = 0$.
Solve the equation:
$x^2 + x - 6 = (x + 3)(x - 2) = 0$, ◄──── These are the points where the curve cuts the x-axis.
so $x = -3$ or $x = 2$.

3 Sketch the curve $x^2 + x - 6$.

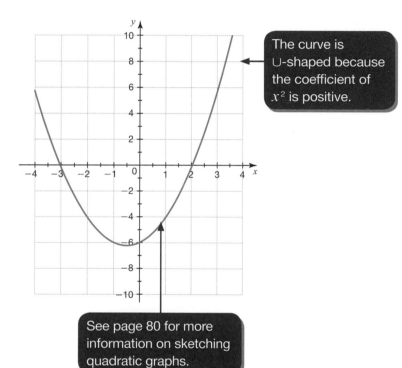

The curve is U-shaped because the coefficient of x^2 is positive.

See page 80 for more information on sketching quadratic graphs.

4 Identify the values of x where $x^2 + x - 6 \geq 0$.
These are the values of x for which y is greater than or equal to 0 – where the graph is above or on the x-axis: $x \leq -3$ and $x \geq 2$.

For the opposite inequality ($x^2 + x - 6 \leq 0$), the values of x would be for the graph on or below the x-axis: $-3 \leq x \leq 2$.

DOIT!

Make up some inequalities, rolling dice to generate the numbers. Solve them.

WORKIT!

a Solve the inequality $x^2 - 7x + 10 < 0$.

$x^2 - 7x + 10 = 0$ gives $(x - 2)(x - 5) = 0$, so the curve intersects the x-axis at $x = 2$ and $x = 5$.

> Work out the solutions to $x^2 - 7x + 10 = 0$ and sketch the curve, marking the roots of the quadratic.

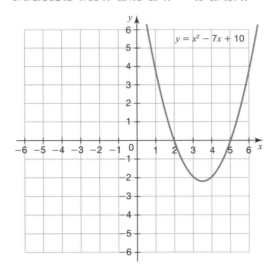

Solution is $2 < x < 5$.

> Identify the values of x for the graph below the x-axis because of the $<$ in the inequality.

b Show the solutions on a number line.

CHECKIT!

1 Solve these inequalities. Show your answers on a number line and using set notation.

a $1 - 2x < -11$

c $\dfrac{x - 5}{3} < 7$

b $2x - 7 \geq 15$

2 Solve these inequalities.

a $2x - 4 > x + 6$

b $4 + x < 6 - 4x$

c $2x + 9 \geq 5(x - 3)$

3 a Illustrate the region represented by the following inequalities on a graph.

$x > -1 \qquad y \leq 2 \qquad x - y < 2$

b List the integer values of x and y that satisfy all of these inequalities.

4 Solve the inequality $x^2 > 3x + 10$.

Problem solving using algebra

Problems involving simultaneous equations

In some questions you will have to form the two simultaneous equations from information given in the question and then solve them.

Problems involving inequalities

In some problems, you will be given a maximum or minimum value (e.g. length, area, volume). You then set up an inequality and solve it.

WORKIT!

The sum of two numbers is 14 and the difference is 2. Find the two numbers.

Let the larger number $= x$ and the smaller number $= y$.

Create the simultaneous equations from the information in the question.

$x + y = 14$ (1)

$x - y = 2$ (2)

Add equations (1) and (2).

$2x = 16$

$x = 8$

$x + y = 14$ (1)

Substitute $x = 8$ into equation (1).

$8 + y = 14$

$y = 6$

Check:

Substitute $x = 8$ and $y = 6$ into equation (2).

$x - y = 2$ (2)

$8 - 6 = 2 = x - y$

The two numbers are 6 and 8.

WORKIT!

This solid cuboid has length $4x$ cm, width $2x$ cm and height x cm, where x is an integer.

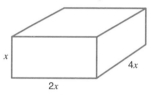

Find an expression for the total surface area in terms of x.

The total surface area of the cuboid is less than or equal to 2800 cm^2.

Show that $x \leq 10$.

Area of top $+$ bottom $= 8x^2 + 8x^2 = 16x^2$

Area of front $+$ back $= 4x^2 + 4x^2 = 8x^2$

Area of the two sides $= 2x^2 + 2x^2 = 4x^2$

Total surface area $= 16x^2 + 8x^2 + 4x^2 = 28x^2$

$28x^2 \leq 2800$

Express the total surface area as an inequality.

$x^2 \leq 100$

$x \leq 10$

$\sqrt{100} = \pm 10$, but it would not make sense for x to be negative when finding the sides of the cuboid.

CHECKIT!

1 Find two numbers such that their sum is 77 and their difference is 25.

2 A number is added to the numerator and denominator of the fraction $\frac{15}{31}$

The resulting fraction is $\frac{5}{6}$

Find the number.

3 The perimeter of a rectangle is 24 cm and its area is 27 cm².

Find the length and width of the rectangle.

Use of functions

A **function** takes a number as **input** and gives another number as **output**.

It does something mathematical to the number you input. Whatever the input it performs the same mathematical operation or operations on it to give the output.

Number machines

We can use a number machine to show the input and the operations performed on the input to give the output. The function 'divide by 3 and then add 1' can be represented by this function machine.

If the input is 12 then the result after the first box will be 12 ÷ 3 = 4, and after the second box it will be 4 + 1 = 5. The final output is 5.

Inverse functions

The **inverse** of a function will produce the input value from the output value.

Suppose the output value of the function is 13. To work out the input value for the function shown above, go backwards through the function machine, reversing the operations in the process: first subtract 1 and then multiply by 3 (i.e. 13 − 1 = 12 and 12 × 3 = 36) to give the input 36.

WORKIT!

A function is represented by the following function machine.

input ⟶ × 3 ⟶ − 4 ⟶ output

a A number is input into the machine.
   The output is used as a new input.
   The second output is 20.
   Work out the number that was the **first** input.

Work backwards through the machine to find the second input.

$20 + 4 = 24$ and $24 ÷ 3 = 8$

Second input = 8

The second input is the first output, so work backwards again with the new figure.

$8 + 4 = 12$ and $12 ÷ 3 = 4$

First input = 4

b When a certain number is input into the function machine the output is the same number. Work out this number.

Let input value = x

Output = $3x − 4$

$3x − 4 = x$

The output ($3x − 4$) is equal to the input (x).

$2x = 4$

$x = 2$

So input value = 2

WORKIT!

A function is represented by the following function machine.

input ⟶ ×___ ⟶ −___ ⟶ output

When the input to the machine is 3, the output is 7.
When the input to the machine is 7, the output is 23.
Work out the numbers in the function machine boxes.

Let the number being multiplied be x and the number being subtracted be y.

$3x - y = 7$ (1) ◄— Use the inputs and outputs to set up two simultaneous equations.

$7x - y = 23$ (2)

$4x = 16$ ◄— Subtract (1) from (2).

$x = 4$

$3(4) - y = 7$ ◄— Substituting $x = 4$ into equation (1).

$y = 5$ ◄— Remember to check the two values by substituting them into equation (2).

Completing the diagram gives:

input ⟶ × 4 ⟶ − 5 ⟶ output

NAILIT!

You could substitute different values until you find the correct ones. However, the numbers may not be integers so it is better to use algebraic methods.

Composite functions

Composite functions involve applying two or more function machines in succession. The output from the first function machine is used as the input for the second function machine.

The order of the function machines is important, as it will usually affect the final output.

WORKIT!

1 The diagram shows two function machines, A and B.

function machine A
input ⟶ ÷ 3 ⟶ + 2 ⟶ output

function machine B
input ⟶ + 1 ⟶ × 3 ⟶ output

The output from function machine A is used as the input for function machine B.

a Find the final output from the two machines when the input is 12.

Machine A: $12 ÷ 3 = 4$ $4 + 2 = 6$ ◄— Find the output from machine A.

Machine B: $6 + 1 = 7$ $7 × 3 = 21$

Final output $= 21$ ◄— Use this as the input for machine B.

b If the final output is 30, find the value of the input to machine A.

Machine B: $30 ÷ 3 = 10$ $10 - 1 = 9$ ◄— Find the input to machine B.

Machine A: $9 - 2 = 7$ $7 × 3 = 21$

Input $= 21$ ◄— Use this as the output from machine A.

Using function notation

Functions are denoted by the letter f, with the variable in brackets: f(x), for example.

This function can be written as $f(x) = \frac{x}{3} + 1$.

When applied to the number 18, it is expressed as $f(18) = \frac{18}{3} + 1 = 7$.

Similarly, the function 'square x and then add x' can be written as $f(x) = x^2 + x$.

When applied to the number 3, it is expressed as $f(3) = 3^2 + 3 = 12$.

> Replace x with 3.

WORKIT!

$f(x) = x^2 - 2x + 1$

Work out

a f(2)

> Substitute $x = 2$ in $x^2 - 2x + 1$.

$f(2) = 2^2 - 2(2) + 1 = 1$

b f(−1)

> Put brackets around negative numbers.

$f(-1) = (-1)^2 - 2(-1) + 1 = 4$

c the values of x for which $f(x) = 0$.

When $f(x) = 0$, $x^2 - 2x + 1 = 0$

$(x - 1)(x - 1) = 0$

So $x = 1$

> This is a quadratic equation that needs to be factorised and then solved.

SNAPIT! Inverse functions

To find the inverse of a function:

1 Let the function equal y.

2 Rearrange the resulting equation so that x is the subject of the equation.

3 Replace x with $f^{-1}(x)$ and replace y with x.

WORKIT!

$f(x) = \frac{x}{3} + 1$. Find $f^{-1}(x)$.

Let $y = \frac{x}{3} + 1$

$3y = x + 3$

> Rearrange to make x the subject. First multiply both sides by 3 to remove the denominator.

$x = 3y - 3$

$f^{-1}(x) = 3x - 3$

> Replace x with $f^{-1}(x)$ and replace y with x.

Composite functions can be denoted by the letters f and g: f for the first function and g for the second, with the variable in brackets.

Consider the following example:

$f(x) = x^2$ and $g(x) = x - 2$

The composite function $fg(x)$ means $f(g(x))$ and is the result of performing the function g first and then the function f.

To find $fg(x)$, replace x by the expression for $g(x)$ in $f(x)$:

$fg(x) = f(g(x)) = f(x - 2) = (x - 2)^2$

> Here, g means 'subtract 2 from it' and f means 'square it'. So fg means 'subtract 2 from it' and then 'square it', i.e. $(x - 2)^2$. gf means 'square it' and then 'subtract 2 from it', i.e. $x^2 - 2$.

WORKIT!

The functions f and g are such that $f(x) = 3(x + 1)$ and $g(x) = \frac{x}{3} + 2$.

Find

a g(12)

$g(12) = \frac{12}{3} + 2 = 4 + 2 = 6$

b $g^{-1}(x)$

Let $y = \frac{x}{3} + 2$

$3y = x + 6$

$x = 3y - 6$

$g^{-1}(x) = 3x - 6$

c $fg(x)$

$fg(x) = 3\left(\left(\frac{x}{3} + 2\right) + 1\right)$

$= 3\left(\frac{x}{3} + 3\right)$

$= x + 9$

DOIT!

Find an inverse function, using both the function machine method and the algebraic method.

CHECKIT!

1 The diagram shows a function machine.

input $\longrightarrow$ ÷ 5 $\longrightarrow$ − 4 $\longrightarrow$ output

a If the input value is x, find an expression for the output value.

b If the output value is x, find an expression for the input value.

2 The diagram shows a function machine.

input $\longrightarrow$ − 1 $\longrightarrow$ find reciprocal $\longrightarrow$ output

a Find the outputs for the following inputs

i 3   **ii** $1\frac{1}{4}$

b Find the input when the output is −3.

c If the input value is x, find an expression for the output value.

3 $f(x) = \frac{1}{x - 1}$ where $x \neq 1$.

Find

a f(0)   **b** $f\left(-\frac{1}{2}\right)$ **c** $f^{-1}(x)$

4 $f(x) = \sqrt{(x^2 - 9)}$ and $g(x) = x + 4$.

Find

a fg(x)   **b** gf(x)   **c** gf(3)

Iterative methods

Iteration means repeatedly performing a function, each time using your previous answer to get the next result. Iterative methods enable equations to be solved that would be hard to solve using other methods.

Using changes of sign of f(x) to locate solutions

If f(x) can take any value between a and b, and there is a change of sign between f(a) and f(b), then a solution of f(x) = 0 lies between a and b.

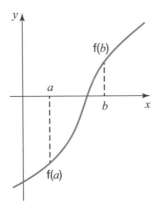

WORKIT!

Show that the equation

$9x^3 - 9x + 1 = 0$

has a solution between 0 and 0.2. ◄—— | a and b are 0 and 0.2, respectively.

Let $f(x) = 9x^3 - 9x + 1$

$f(0) = 1$

$f(0.2) = 9(0.2)^3 - 9(0.2) + 1$

    $= 0.072 - 1.8 + 1$

    $= -0.728$ ◄—— | Insert each value for x into the function and see if there is a sign change between f(0) and f(0.2).

As there is a sign change, there is a solution between 0 and 0.2.

NAILIT!

When you have a function f(x) equal to 0, a solution is found where the graph of the function crosses the x-axis (i.e where the function is equal to 0). The x value at this point is known as a root of the function.

Quadratic functions have two roots (or none if they don't cross the x-axis).

Using decimal search to find a solution

Equations can be solved using a method called **decimal search**. This involves testing different decimal values until you find where the function is equal to 0.

You can do this in simple numerical steps (see top Work it!). A more efficient method is called **binary search** or **interval bisection** (see bottom Work it!).

WORKIT!

A solution to the equation $x^3 - 2x + 8 = 0$ lies between $x = -3$ and $x = -2$. Find this solution to 1 decimal place.

Start at $x = -3$ and go up in steps of 0.1 until the sign changes.

The sign changes between $x = -2.4$ and $x = -2.3$ so the solution lies between these two values.

x	$x^3 - 2x + 8$
-3	-13
-2.9	-10.589
-2.8	-8.352
-2.7	-6.283
-2.6	-4.376
-2.5	-2.625
-2.4	-1.024
-2.3	0.433
-2.35	-0.2779
-2.34	-0.1329
-2.33	-0.0107
-2.32	0.1528

As 0.433 is nearest to 0, $x = -2.3$ is the current best estimate. So start at -2.35 and go up in steps of 0.01 until the sign changes.

The sign changes between $x = -2.33$ and $x = -2.32$ so the solution lies between these two values.

$x = -2.3$ (to 1 d.p.)

Both -2.33 and -2.32 round to -2.3 to 1 decimal place.

WORKIT!

A solution to the equation $x^3 - 2x + 8 = 0$ lies between $x = -3$ and $x = -2$. Use binary search (interval bisection) to find this solution to 1 decimal place.

Find the value of the expression halfway between $x = -2.5$ and $x = -2$. The sign changes between $x = -2.5$ and $x = -2.25$ so the solution lies between these two values.

x	$x^3 - 2x + 8$
-3	-13
-2	4
-2.5	-2.625
-2.25	1.109
-2.35	-0.278

Find the value of the expression halfway between $x = -3$ and $x = -2$.

The sign changes between $x = -2.5$ and $x = -2$ so the solution lies between these two values.

The solution needs to be to 1 decimal place so try $x = -2.35$.

$x = -2.3$ (to 1 d.p.)

The sign changes between $x = -2.35$ and $x = -2.25$ so the solution lies between these values and rounds to -2.3.

Using a recurrence relation to solve an equation

A recurrence relation gives a value in terms of the previous value. Recurrence relations can be used to solve equations, i.e. find a **root**.

> The subscript tags n and $n + 1$ mean that there is a series of values for x, and each solution is used to generate the next one, using this function. We start with $n = 0$.

> a is one of the roots of the equations (i.e. a solution).

> To get x_1, substitute the value for x_0 into the equation. Then substitute x_1 to find x_2, and so on.

> Round the final answer to the required number of decimal places.

WORKIT!

> A root of an equation is a value of x where the function equals 0, i.e. where its graph crosses the x-axis.

The recurrence relation

$$x_{n+1} = \left(\frac{2x_n + 5}{4}\right)^{\frac{1}{3}}$$

with $x_0 = 1.2$, may be used to find a, a root of $4x^3 - 2x - 5 = 0$.

Find and record the values of x_1, x_2, x_3 and x_4. Write down x_4 correct to 5 decimal places and prove that this is equivalent to the root a correct to 5 decimal places.

$x_0 = 1.2$

$x_1 = \left(\frac{2x_0 + 5}{4}\right)^{\frac{1}{3}} = \left(\frac{2(1.2) + 5}{4}\right)^{\frac{1}{3}} = 1.227601026$

$x_2 = \left(\frac{2x_1 + 5}{4}\right)^{\frac{1}{3}} = \left(\frac{2(1.227601026) + 5}{4}\right)^{\frac{1}{3}} = 1.230645994$

$x_3 = \left(\frac{2x_2 + 5}{4}\right)^{\frac{1}{3}} = \left(\frac{2(1.230645994) + 5}{4}\right)^{\frac{1}{3}} = 1.230980996$

$x_4 = \left(\frac{2x_3 + 5}{4}\right)^{\frac{1}{3}} = \left(\frac{2(1.230980996) + 5}{4}\right)^{\frac{1}{3}}$

> Do not round these numbers yet.

$= 1.231017841 = 1.23102$ (to 5 d.p.)

> The root lies on the x-axis where f(x) = 0. Test x values either side of your answer by putting them into the function. If one output is positive and one negative, your answer is in the right range and is correct.

$f(1.231015) = 4(1.231015)^3 - 2(1.231015) - 5 = -0.000119668$

$f(1.231025) = 4(1.231025)^3 - 2(1.231025) - 5 = 0.000042182$

As there is a sign change between these values of f(x), 1.23102 is in the correct range.

DOIT!

> Draw a flowchart for using iteration to find a root to a specified accuracy.

CHECKIT!

1 The equation $3x^3 + 2x^2 - 3 = 0$ has a solution between $x = 0$ and $x = 1$. Find this solution to 1 decimal place.

2 The cubic equation $x^3 - x - 2 = 0$ has a root a between 1 and 2.

The recurrence relation

$$x_{n+1} = (x_n + 2)^{\frac{1}{3}}$$

with $x_0 = 1.5$ can be used to find a.

Calculate x_4, giving your answer correct to 3 decimal places. Prove that this value is also the value of a correct to 3 decimal places.

Equation of a straight line

Straight-line graphs are also called **linear** graphs and have an equation of the form:

$y = mx + c$ ← There is a single y on the left-hand side of the equation.

m is the **gradient** (i.e. the steepness of the line) and c is the **intercept on the y-axis** (where the graph cuts the y-axis).

It is important that in this equation there is only an x term. There are no terms containing x^2, x^3, $\sqrt{x}$, $\frac{1}{x}$, and so on.

This is an equation of a straight line: $y = 2x - 3$.

Comparing this equation with $y = mx + c$, the gradient, m, is 2 and the intercept on the y-axis, c, is -3.

NAILIT!

For an equation to be that of a straight line, you must be able to rearrange it in the form $y = mx + c$ (with no higher power than x).

WORKIT!

Which of these equations are straight lines? Explain your answer.

a $y = 1.4x + 7$

Yes: in the form $y = mx + c$.

b $y = 4x^2 + 1$

No: not in the form $y = mx + c$ as it contains a term in x^2.

c $y = \frac{2}{x}$

No: not in the form $y = mx + c$.

d $5x + 2y = 3$

Yes: can be rearranged to give $2y = -5x + 3$ and this can be divided by 2 to give $y = -\frac{5}{2}x + \frac{3}{2}$, which is in the form $y = mx + c$.

Gradients of straight-line graphs

Gradients have a sign and a value, for example -0.6 and 4. ← The gradient of a line is the steepness of the line.

The sign of the gradient

The gradient of a straight line can be positive (if y increases as x increases), negative (if y decreases as x increases) or zero (if the value of y stays the same as x increases).

SNAPIT! Straight-line graphs

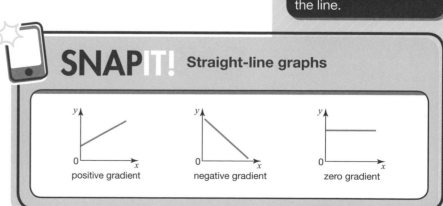

positive gradient   negative gradient   zero gradient

Finding the gradient

To find the size of the gradient, draw a triangle as shown. Then use this relationship for the gradient:

Gradient, $m = \dfrac{\text{change in } y \text{ values}}{\text{change in } x \text{ values}}$

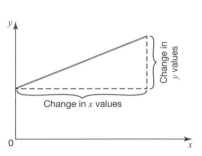

To determine whether the gradient is positive or negative, look at which way the line slopes. Only include the sign if the gradient is negative. If the gradient is positive, say 4, do not write +4.

Finding the gradient of a straight line joining two points

SNAPIT!   Gradient of a line

The gradient of the line joining points (x_1, y_1) and (x_2, y_2) is given by:

Gradient $= \dfrac{y_2 - y_1}{x_2 - x_1}$

From the graph,
$AC = x_2 - x_1$, $BC = y_2 - y_1$ and gradient $= \dfrac{BC}{AC}$.

WORKIT!

Find the gradient of the straight line joining the points A(−3, 2) and B(1, 6).

$Gradient = \dfrac{6 - 2}{1 - (-3)} = \dfrac{4}{4} = 1$

> Take care with the signs. Make sure that you give the x and y coordinates in the same order.

Finding the midpoint of a straight line

NAILIT!

You need to remember this formula as it will not be given in the formula booklet.

SNAPIT!

Midpoint of a line

The midpoint of the line joining the points (x_1, y_1) and (x_2, y_2) is given by:

$\left(\dfrac{x_1 + x_2}{2}, \dfrac{y_1 + y_2}{2} \right)$

WORKIT!

Find the midpoint of the line joining the two points with coordinates (2, 6) and (8, 4).

$\left(\dfrac{2 + 8}{2}, \dfrac{6 + 4}{2} \right) = (5, 5)$

Finding the equation of a straight line

Given the gradient and y-intercept

Given the gradient of a line, m, and the intercept on the y-axis, c, you can write down the equation of the line.

WORKIT!

Find the equation of the line shown.

First find the gradient, m. Try to find a place where the graph cuts through one of the corners of one of the large squares on the grid. Make the triangle as large as possible.

(0, 6) and (8, 12) are convenient points to start and finish the triangle.

$m = \dfrac{\text{change in } y \text{ values}}{\text{change in } x \text{ values}} = \dfrac{6}{8} = \dfrac{3}{4}$ or 0.75

y increases as x increases, so the gradient is positive.

$c = 6$

Look at where the line crosses the y-axis.

$y = \dfrac{3}{4}x + 6$

Substitute for m and c in the equation of a line.

Given the gradient and the coordinates of a point

SNAPIT! Equation of a straight line

The equation of a straight line with gradient m and which passes through a point (x_1, y_1) is given by:

$y - y_1 = m(x - x_1)$

NAILIT!

You need to remember the formula for the equation of a straight line as it will not be given in the formula booklet.

WORKIT!

Find the equation of the straight line with a gradient of 2 that passes through the point (2, 5).

$y - 5 = 2(x - 2)$

Substitute for m, x_1 and y_1.

$y - 5 = 2x - 4$

Expand the brackets and rearrange into the form $y = mx + c$.

$y = 2x + 1$

This is in the form $y = mx + c$, so you can immediately see that $m = 2$ and c (the intercept on the y-axis) is 1.

Parallel and perpendicular lines

Parallel lines

For two lines to be **parallel** to each other, they must have the same gradient.

For example, the equation of the line that is parallel to the line $y = 3x - 2$ and intersects the y-axis at $y = 4$ is:

$y = 3x + 4$ ◄── $m = 3$ and $c = 4$

Perpendicular lines

When two lines are **perpendicular** to each other (i.e. they make an angle of 90°), the product (multiple) of their gradients is −1.

If one line has a gradient m_1 and the other a gradient of m_2 then:

$m_1 m_2 = -1$

For example, if a straight line has gradient $-\frac{1}{3}$, then the gradient of the line perpendicular to this is given by:

$\left(-\frac{1}{3}\right)m_2 = -1$ so gradient $m_2 = 3$.

NAIL IT!

You may be asked to find the equation of a straight line through two points. First find the gradient. Then substitute m and one pair of coordinates into $y = mx + c$.

DO IT!

Write a revision card for the different ways of finding the equation of a straight line.

Reciprocal means 'one over' a value, or that value to the power −1.

NAIL IT!

A quick way to find the gradient of a line that is perpendicular is to write the reciprocal of the gradient of the original line and change the sign $\left(\text{i.e. } -\frac{1}{\text{original gradient}}\right)$.

CHECK IT!

1 A straight line has the equation $2y = 4x - 5$.

   a Write down the gradient of the line.

   A line is drawn perpendicular to this line.

   b Write down the gradient of the perpendicular line.

   The perpendicular line crosses the y-axis at (0, 5).

   c Write down the equation of the line.

2 Find the equation of the line that has gradient 3 and passes through the point (2, 3).

3 A line has a gradient of 2 and passes through the point (−1, 0).

   Find the equation of the line in the form $ay + bx + c = 0$.

4 A straight line passes through the points $A(-2, 0)$ and $B(6, 4)$.

   a Find the gradient of the line AB.

   b The midpoint of AB is M. Find the coordinates of M.

   c A straight line is drawn through point M which is perpendicular to the line AB.

      i Write down the gradient of this line.

      ii Find the equation of this line.

Quadratic graphs

Completing the square

Completing the square can be used to:

- solve quadratic equations
- help draw a quadratic graph by finding turning points.

Completing the square: coefficient of $x^2 = 1$

To complete the square for the quadratic $x^2 - 2x + 7$:

1. Separate the x terms from the constant: $(x^2 - 2x) + 7$.

2. Write the perfect square to give the x^2 and x terms: $(x - 1)^2$.

3. Then subtract a constant to make the identity true: $x^2 - 2x \equiv (x - 1)^2 - 1$.

4. Combine this with the constant in the original quadratic:

$$x^2 - 2x + 7 = (x - 1)^2 - 1 + 7$$
$$= (x - 1)^2 + 6$$

> The **coefficient** of x^2 is the number in front of x^2. If the first term is just x^2, the coefficient is 1.

> This example only used integers, but the method also works with fractions.

Completing the square: coefficient of $x^2 = 2$

To complete the square for the quadratic $2x^2 - 6x + 5$:

1. Remove the coefficient of x^2 to outside a bracket: $2\left[x^2 - 3x + \frac{5}{2}\right]$.

2. Use the method above to complete the square for the contents of the square bracket: $2\left[\left(x - \frac{3}{2}\right)^2 - \frac{9}{4} + \frac{5}{2}\right] = 2\left[\left(x - \frac{3}{2}\right)^2 + \frac{1}{4}\right]$.

3. Multiply the expression found by the number outside: $2\left(x - \frac{3}{2}\right)^2 + \frac{1}{2}$

So, $2x^2 - 6x + 5 = 2\left(x - \frac{3}{2}\right)^2 + \frac{1}{2}$.

Solving a quadratic equation by completing the square

Quadratic equations can be solved by completing the square.

> **NAIL**IT!
>
> You must use completing the square to solve the quadratic equation. You will lose marks if a method is specified in the question and you use a different method.

WORKIT!

a Show that $x^2 + 6x + 4$ may be expressed in the form $(x + 3)^2 - 5$.

$$x^2 + 6x + 4 = (x + 3)^2 - 9 + 4$$
$$= (x + 3)^2 - 5$$

b Use your answer to part a to solve the quadratic equation $x^2 + 6x + 4 = 0$, giving your answers to 2 decimal places.

$$x^2 + 6x + 4 = 0$$
$$(x + 3)^2 - 5 = 0$$
$$(x + 3)^2 = 5$$
$$x + 3 = \pm\sqrt{5}$$
$$x = -3 + \sqrt{5} \text{ or } x = -3 - \sqrt{5}$$
$$x = -0.76 \text{ or } x = -5.24 \text{ (to 2 d.p.)}$$

> You must include both the positive and negative values when you square-root a number, giving two roots (i.e. solutions) rather than just one.

Sketching a quadratic graph

The shape of the graph of the quadratic equation $y = ax^2 + bx + c$ depends on the sign of a. The curve is ∪-shaped if a is positive and ∩-shaped if a is negative.

Finding the point or points where the curve intersects the x-axis

It is important to note that not all quadratic graphs intersect the x-axis.

'Intersect' means 'to cross'.

To find where the graph intersects the x-axis, solve the equation $ax^2 + bx + c = 0$. These values of x are called the **roots** of the equation. Mark the root(s) of x on the x-axis.

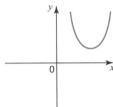

This graph does not intersect the x-axis so there are no roots.

This graph intersects the x-axis in two places so there are two roots.

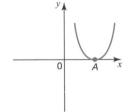

This graph intersects the x-axis in only one place so there is one root.

NAILIT!

Check the question to see if you should give your answer to a certain number of significant figures or decimal places.

Finding turning points by completing the square

The **turning points** on a curve are the points where the gradient of the curve is zero. For a quadratic graph this will be a **maximum** point if the curve is ∩-shaped or a **minimum** point if the curve is ∪-shaped.

When the square has been completed, the equation for the curve will look like:

$y = a(x + p)^2 + q$

When $x = -p$, the value of the bracket is zero and, since the bracket is squared, this is its minimum value (since it cannot be negative). So the minimum value of y is q.

The turning point will be at $(-p, q)$.

The axis of symmetry will be the line $x = -p$.

For example, for $y = 2(x + 2)^2 - 1$, $a = 2$, $p = 2$ and $q = -1$.

The curve will be ∪-shaped with a minimum point at $(-2, -1)$ and an axis of symmetry of $x = -2$.

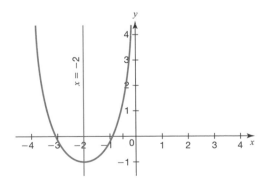

The maximum and minimum points of curves are called turning points as the gradient changes sign either side of the point. In this curve the gradient to the left is negative and to the right is positive. The gradient at the turning point itself is zero.

WORKIT!

a Express $4x^2 - 12x + 9$ in the form $a(x + b)^2 + c$.
Give the values of a, b and c.

$$4x^2 - 12x + 9 = 4\left[x^2 - 3x + \tfrac{9}{4}\right]$$

> Write the coefficient of x^2 as a factor of the expression.

$$= 4\left[\left(x - \tfrac{3}{2}\right)^2 - \tfrac{9}{4} + \tfrac{9}{4}\right]$$

$$= 4\left(x - \tfrac{3}{2}\right)^2$$

$$a = 4, \; b = -\tfrac{3}{2}, \; c = 0$$

> Compare the expression with $a(x + b)^2 + c$.

b Hence sketch the graph of $y = 4x^2 - 12x + 9$, including the coordinates of the turning point.

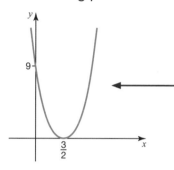

> The coefficient of x^2 is positive, so the graph is ∪-shaped.
>
> At the minimum point, $x = \tfrac{3}{2}$. When $x = \tfrac{3}{2}$, $y = 0$
> The coordinates of the turning point are $\left(\tfrac{3}{2}, 0\right)$.

WORKIT!

This is a quadratic graph.

> The graph is ∩-shaped, so the coefficient of x^2 is negative. Use the roots to write the factors of the quadratic.

a Write down the roots of the graph.

$$x = -2 \text{ and } x = 3$$

b Find the equation of the quadratic graph.

$$-(x + 2)(x - 3) = -(x^2 - x - 6)$$

$$= -x^2 + x + 6$$

The equation is $y = -x^2 + x + 6$.

c Use the graph to estimate the solutions to $-x^2 + x + 6 = 2$.

$$x = -1.55 \text{ and } x = 2.55$$

> Estimate the x values when $y = 2$.

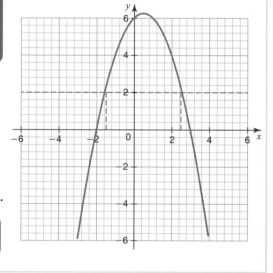

DOIT!

Produce a poster showing the information you need to sketch a quadratic graph.

CHECK**IT!**

1 a Write $2x^2 - 12x + 1$ in the form $a(x + b)^2 + c$, where a, b and c are integers.

b Hence or otherwise, for the graph $y = 2x^2 - 12x + 1$

   i find the coordinates of the turning point

   ii find the roots to 1 decimal place.

c Sketch the graph of $y = 2x^2 - 12x + 1$.

2 Write the equation for each quadratic graph.

a

b

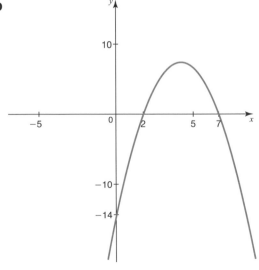

3 a Write $x^2 + 12x - 16$ in the form $(x + a)^2 + b$.

b Write down the coordinates of the turning point of the graph of $y = x^2 + 12x - 16$.

Recognising and sketching graphs of functions

SNAPIT!

When plotting a graph:

1 Label the axes.

2 Mark the origin.

3 Plot each point with a neat X.

4 Join the points with a smooth curve.

5 Include the equation of the line/curve.

6 Do not draw the graph past the x coordinates that were specified.

Drawing cubic graphs

Cubic graphs have an x^3 term and have an equation of the form:

$$y = ax^3 + bx^2 + cx + d$$

where a, b, c and d are constants and where b, c or d could be zero.

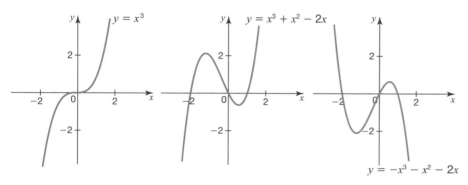

Graph has one root and no turning points.

Graph has two turning points and three roots.

Graph has two turning points and three roots.

WORKIT!

Draw the graph of $y = x^3 + x^2 - 6x$ between $x = -4$ and $x = 3$.

When $x = -4$, $y = (-4)^3 + (-4)^2 - 6(-4) = -24$.

x	-4	-3	-2	-1	0	1	2	3
$y = x^3 + x^2 - 6x$	-24	0	8	6	0	-4	0	18

Draw a table and substitute each value of x into the equation to find the value of y. (The first substitution is shown here.)

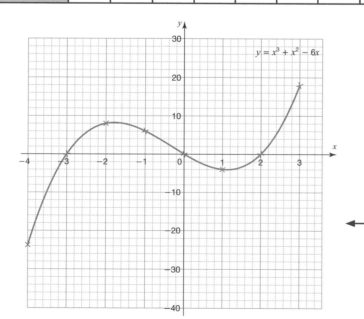

NAILIT!

When substituting into an expression with many terms, it is best to write the calculation down.

Plot the points in the table and join them with a smooth curve.

Drawing exponential graphs

An exponential graph has an equation of the form $y = ka^x$.

To plot the graph of an exponential function, for example $y = 3 \times 1.1^x$, use a table of values.

WORKIT!

A certain fast-growing pond plant can be modelled by:

$N = 5 \times 1.2^t$

where N is the number of strands present after t days.

a How many strands are there at $t = 0$ days?

When $t = 0$, $N = 5 \times 1.2^0 = 5$ ◄

> Any number or letter to the power 0 is always 1, so $1.2^0 = 1$.

b Draw the graph of $N = 5 \times 1.2^t$ for the first 8 days.

Time, t (days)	0	1	2	3	4	5	6	7	8
Number of strands, N	5.0	6.0	7.2	8.6	10.4	12.4	14.9	17.9	21.5

NAILIT!

Often the questions on sketching graphs will give a series of graphs and a series of equations for you to match each equation to the correct graph.

> Draw a table for $t = 0$ to 8. Work out the value of N for each value of t.

> Plot the points in the table and join them with a smooth curve.

NAILIT!

If you are asked to draw a sketch graph, make sure you label the x- and y-axes and the origin and write the equation next to the curve. You may also be asked to mark the coordinates of any points where the curve crosses the axes.

> Exponential functions are ones in the form $y = a^x$. You will only see exponential functions where a is a positive number.

Recognising sketch graphs

You looked at sketching quadratic graphs in the previous section. You also need to be able to recognise and sketch graphs of the following functions.

Reciprocal function   **Cubic function**   **Exponential function**

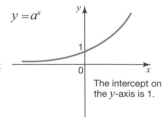

84

The trigonometric functions

You need to know about the three trigonometric functions, **sine**, **cosine** and **tangent** – often abbreviated to sin, cos and tan.

There is more information on the trigonometric ratios on page 158.

The sine graph, $y = \sin x$

The cosine graph, $y = \cos x$

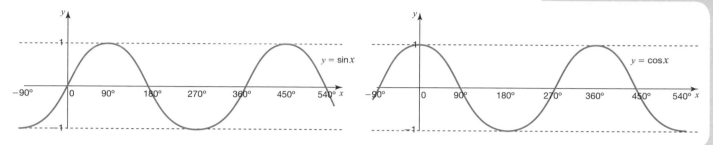

The tangent graph, $y = \tan x$

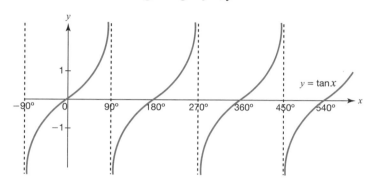

Using the trigonometric graphs to find angles

These graphs can be used to find angles.

WORKIT!

Find all the values of the angle θ in the range $0° \leq \theta \leq 360°$ satisfying $2\sin\theta = 1$.

$2\sin\theta = 1$

$\sin\theta = \frac{1}{2}$

$\theta = \sin^{-1}\left(\frac{1}{2}\right)$

Sin^{-1} means 'angle with a sine of'.

$\theta = 30°$

Also, $\theta = 180 - 30 = 150°$

The line $y = 0.5$ (or $y = \frac{1}{2}$) lets you read off the values for the angle.

Hence $\theta = 30°, 150°$

DOIT!

Type an equation into an internet search engine and find out what the graph looks like.

Make a small change to the equation and search again to see how this changes the graph.

To produce the graph $y = x^3$ you need to enter the equation as $y = x\char`\^3$ as the $\char`\^$ symbol is used to indicate a power or **exponential**.

WORKIT!

The graph shows the curve $y = \sin x$ in the interval $0° \leq x \leq 720°$.

a Write down the coordinates of all the points of intersection with the x-axis.

(0, 0), (180, 0), (360, 0), (540, 0), (720, 0)

b Write down the coordinates of all the turning points for this graph.

(90, 1), (270, −1), (450, 1), (630, −1)

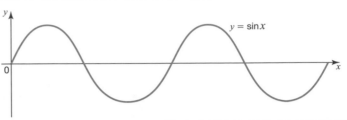

CHECKIT!

1 Here are the equations of six graphs.

a $y = 5x^3$

b $y = \cos x$

c $y = 3^x$

d $y = x^2 - 4x + 4$

e $y = \sin x$

f $y = \dfrac{2}{x}$

Match each equation to one of the graphs A to F.

b One solution to $\tan x = \sqrt{3}$ is $x = 60°$.

Find another solution for x in the range $0° \leq x \leq 360°$.

3 Here are the equations of four graphs. Match each equation to one of the graphs A to G.

a $y = 1 - x$

b $y = 2x - 1$

c $y = x^2 - 3$

d $y = 3 - x^2$

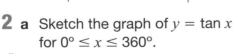

2 a Sketch the graph of $y = \tan x$ for $0° \leq x \leq 360°$.

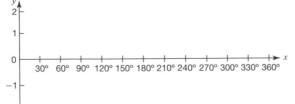

Translations and reflections of functions

Translating the graph of $y = f(x)$

A **translation** is a vertical movement, a horizontal movement or both. If you are given a graph of a function in the form $y = f(x)$, then the graph of a new function may be obtained from the original graph by applying a translation.

$y = f(x)$ translated to $y = f(x + a)$ is a **translation** of $-a$ units **parallel to the x-axis**. The whole curve moves a units to the left. Similarly, $y = f(x - b)$ is a translation of b units to the right.

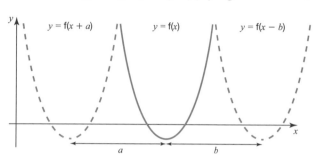

$y = f(x)$ translated to $y = f(x) + a$ is a **translation** of a units **parallel to the y-axis**. If a is positive, the whole graph moves up a units and if a is negative it moves down by a units.

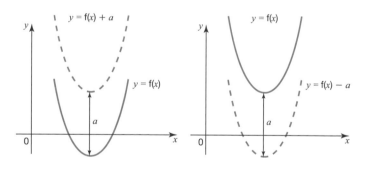

Reflecting the graph of $y = f(x)$

$y = -f(x)$ is a **reflection** of $y = f(x)$ in the **x-axis**.

$y = f(-x)$ is a **reflection** of $y = f(x)$ in the **y-axis**.

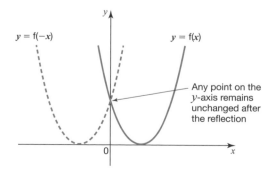

WORKIT!

1   The diagram shows a sketch of the graph $y = f(x)$. The graph passes through the points (1, 0) and (5, 0) and has a turning point at (3, −4).

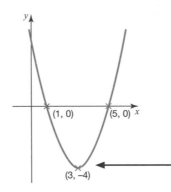

> A turning point is a point on the curve where the gradient is zero. This will occur at any maximum or minimum points.

Sketch the following graphs, using a separate set of axes for each graph. In each case, indicate the coordinates of the turning point and the coordinates of the points of intersection of the graph with the x-axis.

a $y = f(x + 1)$

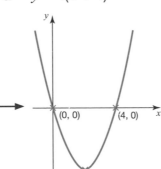

> $y = f(x + 1)$ is a translation of $y = f(x)$ by one unit to the left parallel to the x-axis.

b $y = -f(x)$

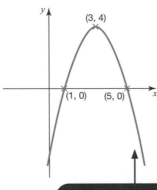

> The negative sign is before f(x), so this is a reflection in the x-axis.

NAILIT!

Remember to add the coordinates of the points of intersection with the x-axis and of the turning point. You would lose marks if you only put the x coordinates of these points on the x-axis.

Combinations of reflections and translations

Translations and reflections can be combined. The order in which they are carried out does not matter.

Here are some combinations of reflections and translations for this graph of $y = f(x)$.

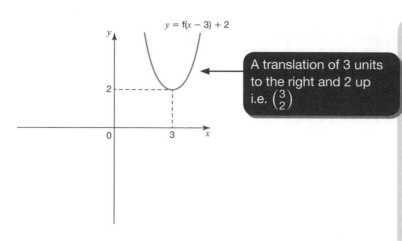

A translation of 3 units to the right and 2 up i.e. $\begin{pmatrix} 3 \\ 2 \end{pmatrix}$

Draw a poster summarising translations and transformations of functions.

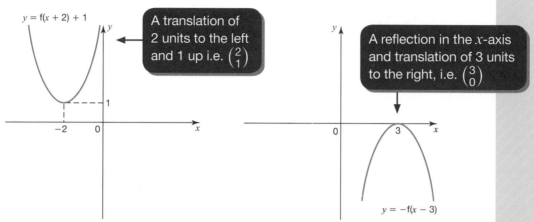

A translation of 2 units to the left and 1 up i.e. $\begin{pmatrix} 2 \\ 1 \end{pmatrix}$

A reflection in the x-axis and translation of 3 units to the right, i.e. $\begin{pmatrix} 3 \\ 0 \end{pmatrix}$

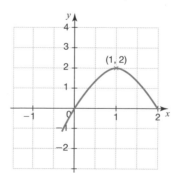

CHECKIT!

1 The diagram shows the graph of $y = f(x)$. The graph has a turning point at (2, 5) and intersects the x-axis at the points (−2, 0) and (6, 0).

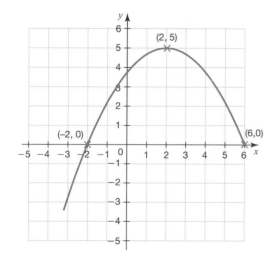

Write down the coordinates of the turning point of the curve with equation

a $y = f(x − 1)$ c $y = −f(x)$

b $y = f(x + 3)$ d $y = f(−x)$

2 The diagram shows the graph of $y = f(x)$. The graph has a maximum point at (1, 2).

Sketch the following graphs, using a separate set of axes for each graph. Mark the coordinates of the turning point on each graph.

a $y = f(x − 1)$ c $y = −f(x − 1)$

b $y = f(x) + 2$ d $y = f(x − 1) + 2$

Equation of a circle and tangent to a circle

Equation of a circle

SNAPIT! Equation of a circle

The equation of a circle with centre the origin (0, 0) and radius r is:

$$x^2 + y^2 = r^2$$

So the equation $x^2 + y^2 = 25$ represents the equation of circle with centre at the origin and with a radius of 5.

WORKIT!

A circle has the equation $x^2 + y^2 = 9$.

a Write down the coordinates of the centre of the circle and give the radius of the circle.

Centre is at (0, 0).

Radius = 3

Comparing with $x^2 + y^2 = r^2$, $r^2 = 9$.

b Prove that the point $P (\sqrt{3}, \sqrt{6})$ lies on the circle.

If these points lie on the circle, then:

$$x^2 + y^2 = (\sqrt{3})^2 + (\sqrt{6})^2$$

$$= 9$$

If the point lies on the circle its coordinates will satisfy the equation of the circle.

This is the same as the right-hand side of the equation of the circle. The point therefore satisfies the equation and so lies on the circle.

Equation of a tangent to a circle

The tangent to a circle is perpendicular to the radius. Using the coordinates of where the tangent touches the circle and the coordinates of the centre of the circle, find the gradient of the radius joining these two points using the formula:

$$\text{Gradient} = \frac{y_2 - y_1}{x_2 - x_1}$$

Use this gradient to work out the gradient of the tangent, as these two lines are perpendicular to each other. If one line has a gradient m_1 and the other a gradient of m_2 then $m_1 m_2 = -1$.

NAILIT!

Remember, perpendicular means 'at right angles to'.

Use the coordinates of the point where the tangent touches the circle and the gradient of the tangent and substitute them into the following formula to give the equation of the tangent:

$$y - y_1 = m(x - x_1)$$

> You learned how to find the equation of a line from the gradient and a point on the line on page 77.

WORKIT!

A circle with equation $x^2 + y^2 = 25$ has centre O and a tangent to the circle at the point $P(3, 4)$.

a Find the gradient of the line OP.

Gradient of $OP = \dfrac{y_2 - y_1}{x_2 - x_1} = \dfrac{4 - 0}{3 - 0} = \dfrac{4}{3}$

b Using your answer to part a, write down the gradient of the tangent to the circle at point P.

As radius OP and tangent are perpendicular the product of their gradients is -1.

Gradient of tangent $= -\dfrac{3}{4}$

c Find the equation of the tangent passing through point P.

$y - y_1 = m(x - x_1)$

$y - 4 = -\dfrac{3}{4}(x - 3)$ ← Use this with $m = -\dfrac{3}{4}$ and $(x_1, y_1) = (3, 4)$.

$4y - 16 = -3x + 9$

$4y = -3x + 25$

DO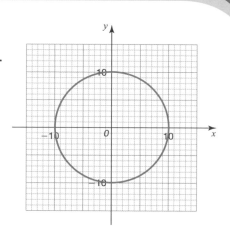IT!

Draw a flowchart for finding the equation of the tangent to a circle.

> The easiest way to find the gradient of the tangent is to invert the gradient of OP (put the fraction upside down) and change the sign.

CHECKIT!

1 A circle has the equation $x^2 + y^2 = 49$.

a Write down the coordinates of the centre of the circle.

b Write down the radius of the circle.

2 The diagram shows a circle with centre at the origin O.

a Write down the equation of the circle.

A tangent to the circle is drawn at $(8, 6)$.

b Find the gradient of the tangent at this point.

c Work out the equation of the tangent at $(8, 6)$.

Real-life graphs

Distance–time graphs

These are the main features of any **distance–time graph**.

- Gradient of a line is the average speed. A steeper line means a higher speed.
- Horizontal sections mean the motion has stopped.
- A negative gradient shows the speed of the return journey.

Below is a distance–time graph for someone's journey to and from a meeting.

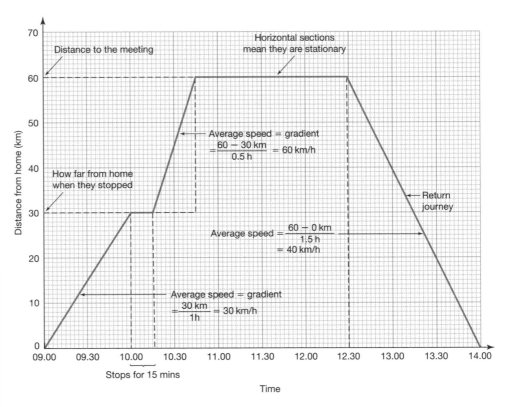

There are lots of things you can tell from this graph and some of them are marked on it. Other things you can tell include:

- The average speed in the first hour is lower than the average speed when they continued after stopping for 15 minutes.
- They stopped at 10 am and waited for 15 minutes.
- The meeting started at 10.45 am and finished at 12.30 pm so it took $1\frac{3}{4}$ hours.

Velocity–time graphs

These are the main features of a **velocity–time graph**.

- The gradient represents the acceleration. A positive gradient represents a positive acceleration and a negative gradient represents a negative acceleration (or deceleration). The units for acceleration are m/s^2.
- Horizontal sections represent constant velocity.
- The area under the graph represents the distance travelled.

DO IT!

Draw and annotate a distance–time graph of a journey to and from a friend's house.

STRETCH IT!

If the velocity–time graph is a curve, you can work out the distance travelled by dividing the area into strips. Look at the diagram shown on the right.

Suppose you have to find the area under the curve between $x = -1$ and $x = 2$.

The area can be approximated to a series of rectangles of equal width. Notice the way the height of each rectangle is adjusted so that the curve cuts the top in two equal parts. You can find the areas of all the rectangles and then add them together to give an estimate of the area under the curve.

You can also find the acceleration. Finding the gradient of a curve is covered on page 121.

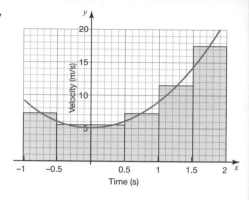

WORKIT!

This velocity–time graph is for a car moving along a straight horizontal road. Its initial velocity is 20 m/s.

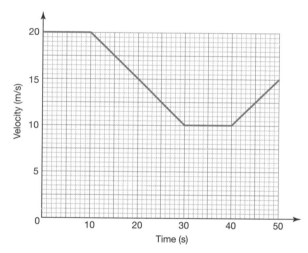

a Find the distance travelled while the car is decelerating.

Distance = area of trapezium ◄──── Base of trapezium is when the car is decelerating, between 10 and 30 seconds.

$= \frac{1}{2}(20 + 10) \times 20$

$= 300\,m$

Sides of trapezium are the velocity before decelerating and the velocity after.

b Find the total distance travelled while travelling at constant speed.

Total distance while travelling at constant speed = $20 \times 10 + 10 \times 10 = 300\,m$ ◄──── Find areas of the rectangles where the car is travelling at constant speed.

c During the last stage of the motion the car accelerates. Calculate the acceleration.

Acceleration = gradient = $\frac{15 - 10}{10} = 0.5\,m/s^2$

d Calculate the total distance travelled during the motion described by the graph.

Distance travelled while accelerating = $\frac{1}{2}(10 + 15) \times 10 = 125\,m$

From a, b and d, total distance travelled = $300 + 300 + 125$

$= 725\,m$ ◄──── The total distance is the total area under the velocity–time graph.

DO IT!

Use the graph shown and the rectangles to estimate the area under the curve between $x = -1$ and $x = 2$.

Area of trapezium $= \frac{1}{2}$ (sum of the two parallel sides) × distance between them. Alternatively, divide the shape into a triangle and a rectangle and add the two areas together. Note you will not be given the formula for the area of a trapezium.

Graphs in finance

Graphs are used in finance to show costs.

This graph shows how the cost of water varies with the volume used.

These are the key features of the graph:

- The intercept on the y-axis shows that there is a fixed cost when no water is used.

- Reading off the graph, the fixed charge is £34.

- The gradient of the graph is the cost per m³ of the water.
  The line goes through (50, 160) and (0, 34) so these points can be used to work out the cost per m³.

  Cost per m³ $= \frac{160 - 34}{50 - 0} = \frac{126}{50}$

  $= £2.52$

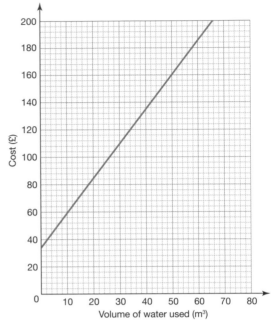

CHECKIT!

1 The distance–time graph represents a cycle ride.

a Find the average speed for the first half-hour.

b At 08.30, the cyclists stopped for a break. How long was this break in hours?

c Find the average speed for the return journey.

2 A particle starts from rest and travels in a straight line. It accelerates uniformly for 2 s, then travels at a constant velocity of u m/s for 10 s, before decelerating uniformly to rest in 3 s. The total distance travelled by the particle is 50 m.

a Draw a velocity–time graph to show the motion of the particle.

b Find the value of u.

c Find the magnitude of the deceleration.

Generating sequences

Arithmetic progression or geometric sequence

An **arithmetic progression** (sometimes called a linear sequence) is a sequence of numbers with a constant difference (positive or negative) between each number and the one before. Add that same amount to give the next number in the sequence.

1, 3, 5, 7,…   Add 2 to get the next term.

13, 18, 23, 28,…   Add 5 to get the next term.

10, 6, 2, −2, −6,…   Add −4 (i.e. subtract 4) to get the next term.

WORKIT!

The first four terms of a sequence are 33, 30, 27, 24,…

a Write down the next two terms.

   21, 18 ◄

b What is the rule for continuing this sequence?

   Subtract 3

   The sequence goes down by 3 each time.

c What is the first negative number in this sequence?

   18, 15, 12, 9, 6, 3, 0, −3

   The first negative term is −3.

A **geometric sequence** (sometimes called a geometric progression) is a sequence of numbers with a constant ratio between each number and the one before. Multiply each number in the sequence by the same amount to give the next number in the sequence.

1, 3, 9, 27, 81,…   Multiply by 3 to get the next term.

$\frac{1}{2}, \frac{1}{4}, \frac{1}{8}, \frac{1}{16},…$ Multiply by $\frac{1}{2}$ to get the next term.

1, √3, 3, 3√3,…   Multiply by √3 to get the next term.

Other common sequences

SNAPIT! Common sequences

Sequence	Name	Rule
1, 4, 9, 16, 25, 36,…	Square numbers	Square the number of each term
1, 8, 27, 64, 125, 216,…	Cube numbers	Cube the number of each term
1, 3, 6, 10, 15, 21,…	Triangular numbers	As shown in the diagram on page 96
1, 1, 2, 3, 5, 8,…	Fibonacci numbers	Add together the previous two terms ◄

DOIT!

Summarise the difference between an arithmetic and a geometric sequence in 21 words.

A Fibonacci sequence usually starts 1, 1, but can start with any two numbers.

Triangular numbers

1   3   6   10   15

NAILIT!

Remember these common sequences, as they often come up in exam questions.

WORKIT!

Write down the next two terms in these sequences.

a $\frac{1}{9}, \frac{1}{3}, 1, 3,...$

9, 27

b 7, 5, 3, 1,...

−1, −3

c 1, 8, 27,...

64, 125 ← The cube numbers.

d 1, 1, 2, 3, 5,...

8, 13 ← Fibonacci sequence.

A geometric sequence with a ratio of 3.

An arithmetic sequence with a common difference of −2.

NAILIT!

Sometimes you will be given the rule and a term, and be asked to find the previous term. You need to use the inverse operations from the rule: for example, 'multiply by 2 and subtract 3' would become 'add 3 and divide by 2'. Note that you need to do the opposite operations and in the **opposite order**.

✓ CHECKIT!

1 Write down the next term of each of these sequences

a 5, 8, 11, 14,...

b 2.2, 2.4, 2.6, 2.8,...

c 0, −3, −6, −9,...

d 2, 12, 72,...

e $\frac{1}{6}, \frac{1}{12}, \frac{1}{24},...$

f $\frac{1}{2}, -\frac{1}{4}, \frac{1}{8},...$

2 The term-to-term rule for a sequence is square and add one. The first term of the sequence is −4. Find the 2nd and 3rd terms.

3 The term-to-term rule for a sequence is double and add one. The third term of the sequence is 12. Find the 1st and 2nd terms.

The *n*th term

The *n*th term of a sequence is a formula that gives the value of any term depending on its position in the sequence. For example, to find the 3rd term, put 3 into the formula.

You can find the *n*th term for some sequences 'by inspection' (i.e. by looking at the terms and their position in the sequence). For example:

> The number on the top of the fraction is the term number (*n*) each time.

- the sequence $\frac{1}{2}, \frac{2}{3}, \frac{3}{4}, \dots$ has *n*th term $\frac{n}{n+1}$
- the sequence $1, \sqrt{3}, 3, 3\sqrt{3}, \dots$ has *n*th term $(\sqrt{3})^{n-1}$

> You can quickly see these terms are $\sqrt{3}$ multiplied by itself (powers of $\sqrt{3}$). The power for the 1st term is 0, not 1. So for the *n*th term the power is $n-1$.

The *n*th term of a linear sequence

To find the *n*th term of the linear sequence: 8, 13, 18, 23,...

n	1	2	3	4
Term	8	13	18	23
Difference		5	5	5

> Write the position of the term above the term.

> Find the difference between terms.

As there is a constant difference of 5 the first part of the formula will be 5*n*.

5*n*	5	10	15	20
Term − 5*n*	3	3	3	3

> Work out the difference between each term and 5*n*.

The *n*th term is 5*n* + 3.

> Sometimes you may have to subtract a value instead of add.

WORKIT!

Here are the first four terms of a sequence 2, 5, 8, 11.

a Write an expression, in terms of *n*, for the *n*th term of the sequence.

n	1	2	3	4
Term	2	5	8	11
Difference		3	3	3
3*n*	3	6	9	12
Term − 3*n*	−1	−1	−1	−1

*n*th term = 3*n* − 1

NAILIT!

Another way of writing 'the *n*th term' is u_n, so in this case
$u_n = 3n - 1$.

b Find the 10th term in the sequence.

10th term = 3 × 10 − 1 = 29

> Substitute *n* = 10 into the formula for the 10th term.

c Is 200 a term of this sequence? Explain your answer.

$3n - 1 = 200$

$3n = 201$

$n = 67$

> Put the *n*th term formula equal to the term. If *n* works out as an integer, it is a number in the sequence.

As *n* is a whole number, 200 is a term in the sequence.

The nth term of a quadratic sequence

In a quadratic sequence the nth term includes a term in n^2.

To find the nth term for the sequence 4, 9, 16, 25,… first find the first and second differences between the terms:

Term		4		9		16		25
First difference			5		7		9	
Second difference				2		2		

> The second difference is the difference between successive first differences.

A constant second difference means that there must be an n^2 term. To find the number in front of n^2, divide the second difference by 2 (in this case, $2 \div 2 = 1$).

So the formula starts n^2. So we can summarise what we know:

n	1	2	3	4
Term	4	9	16	25
n^2	1	4	9	16
Term $- n^2$	3	5	7	9

> We know each term contains n^2. So work out the difference between each term and n^2.

Use the differences between this set of values to work out the linear part of the sequence (the term in n).

> A difference of 2 means the linear part will start with $2n$.

Difference		2		2		2	
$2n$	2		4		6		8
Term $- n^2 - 2n$	1		1		1		1

Combining the terms gives nth term $= n^2 + 2n + 1$.

DO IT!

> Draw a flowchart for finding the nth term of a sequence. The chart needs to work with both arithmetic and quadratic sequences.

✓ CHECK IT!

1 The nth term of a sequence is $50 - 3n$.

   a Write down the first three terms of this sequence.

   b Does the sequence contain the number 34?

   c Work out the value of the first term of this sequence that is negative.

2 The nth term of a sequence is given by 2×3^n.

   a Write down the first four terms of this sequence.

   b All the terms in this sequence are multiples of 6. Explain why.

3 Write the nth term of these sequences:

   a $1, \dfrac{1}{2}, \dfrac{1}{3}, \dfrac{1}{4},\dots$

   b $\dfrac{1}{\sqrt{2}}, \dfrac{1}{2}, \dfrac{1}{2\sqrt{2}}, \dfrac{1}{4},\dots$

4 The first four terms of an arithmetic sequence are -1, 1, 3, 5.

   a Work out an expression for the nth term of this sequence.

   b The xth term of this sequence has a value of 59. Find the value of x.

5 A sequence has the following first four terms:

   4, 17, 38, 67

   Find a formula for the nth term of this sequence.

Arguments and proofs

Arguments and proofs can crop up in almost any topic. As a result of this there is no set way of tackling them. The main thing to remember is that you have to sometimes put letters for values, angles, sides, and so on. Sometimes there are several ways to solve the problem.

WORKIT!

Here are four expressions

$x - 1$ $3x - 3$ $x^2 - 4$ $2(x + 1)$

x is a positive integer. Explain which expression has a value which is

a always even

2 is a factor of $2(x + 1)$.

So $2(x + 1)$ is always even.

b always a multiple of 3

$3x - 3 = 3(x - 1)$, so 3 is a factor.

So $3(x - 1)$ is always a multiple of 3.

c never zero.

When $x = 1$, $x - 1 = 0$

When $x = 1$, $3x - 3 = 0$

When $x = 2$, $x^2 - 4 = 0$ ◄—— Sometimes you can show something is **not** true by giving one **counter-example**.

$2(x + 1)$ is always greater than zero, because x is always greater than 0.

So $2(x + 1)$ is never zero.

WORKIT!

Prove that $(2x - 1)^2 - (x - 2)^2$ is a multiple of 3 for all integer values of x. ◄—— Expand the brackets.

$(2x - 1)^2 - (x - 2)^2 = 4x^2 - 4x + 1 - (x^2 - 4x + 4)$

$= 4x^2 - 4x + 1 - x^2 + 4x - 4$

$= 3x^2 - 3$

$= 3(x^2 - 1)$ ◄—— Factorise the expression.

The 3 outside the brackets shows that the result is a multiple of 3 for all integer values of x.

NAIL IT!

If you are asked to prove something for even and/or odd numbers, use an expression of the form $2n$ for even numbers and $2n + 1$ for odd numbers.

WORKIT!

Prove that the product of two odd numbers always gives an answer that is odd.

Let two odd numbers be $2m + 1$ and $2n + 1$. ← | $2n$ must be even, so $2n + 1$ must be odd.

> Multiply the two expressions to find the product.

$(2m + 1)(2n + 1) = 4mn + 2m + 2n + 1$

$= 2(2mn + m + n) + 1$

$2(2mn + m + n)$ must be even, so $2(2mn + m + n) + 1$ must be odd.

> Always state what you have been asked to prove at the end of your proof.

Therefore the product of two odd numbers is always an odd number.

✓ CHECK IT!

1 Prove whether each of these statements is true or false.

   a If n is a positive integer, $2n + 1$ is always an odd number.

   b If $x^2 - 9 = 0$, the only value x can be is 3.

   c If $n > 0$, n^2 is always an integer.

   d If $n > 0$, n^2 is always greater than n.

2 Prove that the sum of four consecutive numbers is always even.

3 Prove that the sum of three consecutive integers is always a multiple of 3.

4 a and b are both integers and $a > b$. Prove whether each of these statements is true or false.

   a $\dfrac{a}{b} < 1$

   b $b^2 > a^2$

   c $\sqrt{a} > 1$

1 Multiply out the brackets and simplify where possible

  a $-3(3x - 4)$

  b $4x + 3(x + 2) - (x + 2)$

  c $(x + 3)(2x - 1)(3x + 5)$

2 a Factorise the expression $2x^2 + 7x - 4$.

  b Solve the equation $2x^2 + 7x - 4 = 0$.

3 Simplify

  a $(2x^2y)^3$ **c** $\dfrac{15a^3b}{3a^3b^2}$

  b $2x^{-3} \times 3x^4$

4 Solve the simultaneous equations

  $3x + 2y = 8$

  $5x + y = 11$

5 a Show that $\dfrac{3}{x + 7} = \dfrac{2 - x}{x + 1}$ can be written as $x^2 + 8x - 11 = 0$.

  b Hence solve the equation $\dfrac{3}{x + 7} = \dfrac{2 - x}{x + 1}$, giving your answers to 2 decimal places.

6 Make x the subject of $\dfrac{3y - x}{z} = ax + 2$.

7 The function f is such that $f(x) = \frac{x}{3} + 5$.

  a Find $f^{-1}(x)$.

  b The function g is such that $g(x) = 2x^2 + k$, where k is a constant.

    Given that $fg(2) = 10$, work out the value of k.

8 The nth term of a sequence is $30 - 4n$.

  a Write down the first three terms of this sequence.

  b Work out the value of the first term of this sequence that is negative.

9 A circle has the equation $x^2 + y^2 = 21$. Determine whether the point (4, 3) lies inside or outside this circle.

10 Simplify fully $(\sqrt{x} + \sqrt{9y})(\sqrt{x} - 3\sqrt{y})$.

11 a Write $2x^2 + 8x + 1$ in the form $a(x + b)^2 + c$, where a, b and c are integers.

  b Hence or otherwise, for the graph $y = 2x^2 + 8x + 1$:

    i find the coordinates of the turning point

    ii find the roots to 1 decimal place.

  c Sketch the graph of $y = 2x^2 + 8x + 1$.

12 The perimeters of rectangle $ABCD$ and triangle EFG are the same.

  All measurements are in centimetres.

  Work out the area of the triangle.

13 $A(-5, 2)$, $B(-2, -2)$, $C(2, 1)$ and $D(-1, k)$ are the vertices of a square.

  Find the equation of the diagonal BD.

14 Solve algebraically the simultaneous equations

  $x^2 + y^2 = 4$

  $2y - x = 2$

Ratio, proportion and rates of change

Introduction to ratios

Writing ratios

A **ratio** is a comparison between quantities. For example, the ratio of two quantities a and b can be written as $a:b$ or as the fraction $\frac{a}{b}$.

Ratios are like fractions – cancel them if possible.

Cancel ratios such as $32:48$ by dividing both numbers by the highest number that divides exactly into both. Dividing by 16 simplifies the ratio to $2:3$.

This is the highest common factor.

The quantities must be in the same units before the ratio is cancelled. If the ratios are in different units (e.g. $1\,kg:200\,g$):

1. Convert both to the smaller unit: $1000\,g:200\,g$.

2. Remove the units: $1000:200$.

3. Cancel the resulting ratio if possible: $5:1$.

WORKIT!

1. On a map a distance of $2\,cm$ represents $5\,km$. Write the ratio $2\,cm:5\,km$ in the form $1:n$.

First make the units the same.

$$2\,cm:5\,km = 2:500\,000 = 1:250\,000$$

2. Reduce the ratio $34:6$ to the form $1:n$ where n is a number correct to 2 significant figures.

To change the 34 into 1, divide it by 34.
To keep the ratio the same, divide the other side by 34 too.
Round the value of n to to 2 s.f.

$$34:6 = \frac{34}{34}:\frac{6}{34} = 1:0.1764... = 1:0.18$$

Dividing a quantity into two parts in a certain ratio

SNAPIT! Dividing a quantity in a certain ratio

To divide a quantity in a certain ratio:

1. Add the numbers in the ratio together.

2. Divide this number into the quantity you are dividing up. This gives you what one part represents.

3. Multiply the one part by each number in the ratio in turn.

4. Check that the numbers you get add up to give the original quantity.

NAILIT!

Always check to see if you have to give the value of all the parts or just one of the parts. Always give the answer in the form requested by the question.

WORKIT!

1 £180 is divided in the ratio of 4:5. Find the value of each portion.

Total number of parts = 4 + 5 = 9

1 part = $\frac{£180}{9}$ = £20

So the portions are 4 × 20 = £80 and 5 × 20 = £100

> Check your answer by adding together the quantities: 80 + 100 = 180.

2 Divide 250 kg in the ratio 0.4:2.8.

0.4 : 2.8 = 4 : 28 = 1 : 7

> Convert the ratio to whole numbers by multiplying by 10 and then cancelling.

Total number of parts = 1 + 7 = 8

1 part = $\frac{250}{8}$ = 31.25 kg

So the portions are 1 × 31.25 = 31.25 kg and 7 × 31.25 = 218.75 kg

Finding the total being divided in a certain ratio

Sometimes you are given one of the values in the ratio and asked to find the total.

NAILIT!

The method is the same for a three-part ratio. Dividing a quantity in the ratio 2:5:7, the total number of parts is 2 + 5 + 7 = 14.

WORKIT!

1 The ratio of the number of boys to girls in a choir is 4:5. There are 16 boys in the choir. How many children are in the choir?

4 parts = 16 children so 1 part = 4 children

Total number of parts 4 + 5 = 9 parts

Total number of children = 9 × 4 = 36

2 A bag contains red and blue balls. The ratio of red balls to blue balls is 2:3. There are ten more blue balls than red balls. Calculate the total number of balls in the bag.

> You need to use algebra to solve this question, as you do not know how many red or blue balls there are.

Let number of red balls = x, so the number of blue balls = x + 10

$\frac{\text{number of red balls}}{\text{number of blue balls}} = \frac{2}{3}$

> Remember that ratios can also be written as fractions.

$\frac{x}{x + 10} = \frac{2}{3}$

$3x = 2(x + 10)$

$3x = 2x + 20$

$x = 20$

Total number of balls = x + x + 10 = 50

3 In a vet's practice, the ratio of the number of dogs seen to the number of cats seen in a week is 5:3.

50% of the dogs are over the age of 5.

60% of the cats are over the age of 5.

> This is a more challenging question. With some questions you have to put in your own numbers.

What percentages of all the dogs and all the cats seen are over the age of 5?

Total number of parts = 5 + 3 = 8

Assume there are 80 dogs and cats in total. ← Pick a number that is divisible by 8, e.g. 80.

Dividing this in the ratio 5:3 will give 50 dogs and 30 cats.

Number of dogs over 5 years old = 50% of 50 = 25

Number of cats over 5 years old = 60% of 30 = 18

Number of dogs and cats over 5 years old = 25 + 18 = 43

Percentage of dogs and cats over 5 years old = $\frac{43}{80} \times 100 = 53.75\%$

DOIT!

Produce a revision card for each type of ratio problem described here.

CHECKIT!

1 Simplify these ratios.

  a 2:6

  b 25:60

  c 1.6:3.6

2 Express these ratios as simply as possible, without units.

  a 250 g:2 kg

  b 25 m:250 mm

  c 2 cl:1 l

3 Share £400 in the ratio 3.5:2.1.

4 The ratio of the number of girls to the number of boys in a gym is 4:3. There are 180 girls in the gym. Calculate how many children belong to the gym.

5 Three daughters are aged 21, 25 and 29 years. They are left £150 000 in a will to be divided between them in the ratio of their ages.

Work out how much money the youngest daughter will receive.

6 The number of male and female guests staying at a hotel is in the ratio 5:2.

40% of the male guests are over the age of 40.

30% of the female guests are over the age of 40.

Work out the percentage of all guests under the age of 40.

7 A money box contains 10 p and 20 p coins. The ratio of 10 p coins to 20 p coins is 5:7. There are six more 20 p coins than 10 p coins.

Calculate the total amount in the money box.

8 In a box of marbles, there are:

  • two times as many blue marbles as red marbles

  • five times as many red marbles as yellow marbles.

Work out the ratio of blue marbles to red marbles to yellow marbles.

Scale diagrams and maps

On a scale diagram or a map, all the dimensions have been reduced by the same proportion.

For example, a scale drawing of a garden design with a scale of 1:50 means that a distance of 1 in any unit on the drawing represents an actual distance 50 times larger. So a pond with a diameter of 5 cm on the drawing would have an actual diameter of 50 × 5 cm = 250 cm (or 2.5 m).

NAILIT!

Always change any measurements into the same units.

WORKIT!

A map has a scale of 1:50000. A distance on the map between two villages is 6 cm. What is the actual distance? Give your answer in km.

1 cm on the map is 50000 cm actual distance.

6 cm on the map is 6 × 50000 = 300000 cm actual distance.

$300000 \text{ cm} = \frac{300000}{100} \text{ m} = 3000 \text{ m}$ ◄─── Convert 300000 cm to km.

3000 m = 3 km

Actual distance = 3 km

NAILIT!

Make sure your answer is clearly stated at the end. Sometimes it is best to write a sentence.

WORKIT!

An artist is producing a scale drawing of a horse that is 1.75 m tall. The artist has decided to use a scale of 1:10. How tall will the horse be on the drawing? Give your answer in centimetres.

1.75 m = 175 cm

Height of horse on drawing $= \frac{1}{10} \times 175 = 17.5$ cm ◄─── The scale of 1:10 means that the height of the horse needs to be $\frac{1}{10}$ of the horse's actual height.

DOIT!

Work out the actual distance represented by 4 cm on maps at scales 1:2500, 1:25000 and 1:500000.

✓ CHECKIT!

1 A map has a scale of 1:500000. The actual distance between two cities is 150 km.

How far apart are the cities on the map? Give your answer in cm.

2 The scale diagram shows an island with a port and two ships *A* and *B* out at sea.

a The actual distance between ship *A* and the port is 10 km. Calculate the scale of the map.

b Work out the actual distance between ships *A* and *B*.

Percentage problems

One quantity as a percentage of another

First express the quantities as a ratio in fraction form and then convert this to a percentage by multiplying by 100.

SNAPIT!

A as a percentage of *B*

Quantity A expressed as a percentage of a quantity B is:

$\frac{A}{B} \times 100\%$

WORKIT!

The theoretical yield of a certain product in a reaction is 0.458 g. The actual yield in an experiment was 0.412 g. Work out the actual yield as a percentage of the theoretical yield. Give your answer to 3 significant figures.

$\frac{actual\ yield}{theoretical\ yield} \times 100 = \frac{0.412}{0.458} \times 100$

$= 89.9563\%$

$= 90.0\%$ (to 3 s.f.)

> The actual yield is the numerator and the theoretical yield the denominator. Multiply the fraction by 100 to convert it to a percentage.

Percentage change

Percentage change is worked out in a similar way:

$$\text{Percentage change} = \frac{\text{change}}{\text{original value}} \times 100$$

SNAPIT!   Percentage increase/decrease

The change can be an increase or a decrease, so:

$$\text{Percentage increase} = \frac{\text{increase}}{\text{original value}} \times 100$$

$$\text{Percentage decrease} = \frac{\text{decrease}}{\text{original value}} \times 100$$

> Increase = final value − initial value
> Decrease = initial value − final value

WORKIT!

After 1 year a new car originally costing £15 000 is worth £12 000. What is the percentage decrease?

Decrease = initial value − final value = 15 000 − 12 000 = £3000

Percentage decrease = $\frac{decrease}{original\ value} \times 100 = \frac{3000}{15000} \times 100 = 20\%$

Adding a percentage onto a quantity

In many situations a certain percentage is added onto a quantity. For example, the price of an article could increase by a certain percentage or you may need to find the price of an article after the VAT has been added on.

> You could work out 20% of 13500 and then add this on, but there is a quicker way if you can use a calculator.

WORKIT!

> You are adding 20% to the original amount (i.e. 100%), so you need to find 120% of the original amount, which is the same as multiplying by 1.2.

1 A car costs £13500 before VAT. Find the cost of the car after VAT of 20% has been added.

> Multiply the original cost by the multiplier.

$120\% = \frac{120}{100} = 1.2$ ← Work out the multiplier.

Price of car including VAT at 20% = 13500 × 1.2 = £16200

2 A company's sales in one year were £21200. The sales the next year increased by 140%. Work out the sales in the second year.

> 100% + 140% = 240%.

Multiplier = $240\% = \frac{240}{100} = 2.4$

Second year's sales = 21200 × 2.4 = £50880

Subtracting a percentage from a quantity

Some things go down in price. For example, the value of most cars goes down with age (called depreciation); goods in a sale are often reduced by a certain percentage.

WORKIT!

In a sale, all goods are reduced by 35%. The original price of a pair of shoes was £124. Work out the sale price for the shoes.

$\frac{35}{100} = 0.35$ and $1 - 0.35 = 0.65$ ←

> Work out the multiplier by subtracting the multiplier for 35% from 1.

Sale price = 124 × 0.65 = £80.60

WORKIT!

Jarinda is buying a computer from a shop that offers interest-free credit.

The computer costs £450. She has to pay a deposit of 20% and then a fixed monthly payment over 2 years. Work out how much the fixed monthly payment will be, without using a calculator.

10% of £450 = £45 ←

> Divide by 10 to find 10%.

20% of £450 = £90 ←

> Double 10% to find 20%.

Deposit = £90

Remainder to be paid over 2 years = 450 − 90 = £360

Monthly payment = $\frac{360}{24} = £15$

NAILIT!

Other terms that you may need to know include:

- **credit**: money added to a bank account
- **debit**: money taken out of a bank account
- **bank statement**: list of credits (wages, interest) and debits (payments) and a balance, which can be positive or negative (if you are overdrawn).

NAILIT!

Use the easiest method that works. Here, you don't need to work out 1%, only 10%, as 20% is a simple multiple of that.

Finding the original value

Sometimes you are given a value after a certain percentage is added and need to find the original value. For example, most shops show a price with the 20% VAT included so you need to calculate the original price before the VAT was added.

> You cannot get the correct answer by finding 20% of £18000 and then subtracting it from £18000. This is because the 20% is of the original price, which is a smaller amount.

> The price before the VAT was 100%, so the price including VAT is 120% of the original price.

> Alternatively, 18000 ÷ 1.2 = 15000.

WORKIT!

1 A car costs £18000 including VAT at 20%. Find the cost of the car before the VAT was added.

120% of original price = 18000

10% of original price = $\frac{18\ 000}{12}$

= 1500

100% of original price =
1500 × 10 = 15000

Original price = £15000

> The price after the discount is (100 − 15) = 85% of the original price.

2 A tablet computer costs £204 after a discount of 15% in a sale. Calculate the original price of the computer before the discount.

85% of original price = 204

5% of original price = $\frac{204}{17}$ = 12

100% of original price =
12 × 20 = 240

Original price = £240

> You can spot that 5 × 17 = 85.

Simple interest

If you put money into a savings account, you will be paid a percentage of the amount in interest. Simple interest means that the amount paid is not added to the original amount and re-invested (for example because it is withdrawn) so the original amount stays the same.

To work out simple interest:

1 Work out the interest that would be paid at the end of one year.

2 Multiply this amount by the number of years the money is invested to give the total interest paid.

WORKIT!

£10000 is invested in an account paying simple interest of 4.5% each year. Find the total amount of interest earned over 5 years.

Amount of interest in one year = 4.5%

of £10000 = $\frac{4.5}{100}$ × 10000 = £450

Total interest paid over 5 years =
5 × £450 = £2250

NAIL IT!

When you read the question, decide whether your answer should be larger or smaller than the original value – then check it is when you've finished.

DO IT!

Write down how to find the multiplier for percentage increase and decrease questions.

CHECKIT!

1 In a batch of 300 eggs, 8 eggs were bad. Work out the number of bad eggs as a percentage of the total number of eggs. Give your answer to 2 decimal places.

2 A football manager earns £600000 per year. On the promotion of his team to the Premier League, his earnings increase to £1.1 million per year. Find the percentage increase in his pay. Give your answer to 1 decimal place.

3 Jasmeen earns a salary of £38000 per year in her job as a nurse. She is awarded a pay rise of 3.5%. Find her new salary.

4 In a sale, a mobile phone is reduced by 18%. The sale price is £291.92. Work out the original price.

5 £12000 is invested at an interest rate of 3.5% for 6 years. If simple interest is paid, find the total amount of interest paid over the 6 years.

Direct and inverse proportion

Direct proportion

$y \propto x$ means that y is **directly proportional** to x: as x increases so does y. For example, doubling x will double y, and halving x with halve y.

The proportional sign can be replaced by an equals sign provided a constant, usually called k is included. k is called the **constant of proportionality** and can be found by substituting known values into the equation for x and y.

For direct proportion, $y = kx$.

If y is directly proportional to x^2 we write $y = kx^2$.

WORKIT!

Write the statement a is directly proportional to the square of b mathematically.

Substitute for a and b to find the value of k.

Substitute k back to find the formula relating a and b.

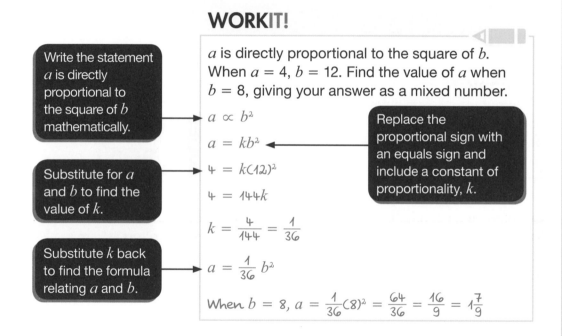

a is directly proportional to the square of b. When $a = 4$, $b = 12$. Find the value of a when $b = 8$, giving your answer as a mixed number.

$a \propto b^2$

$a = kb^2$

Replace the proportional sign with an equals sign and include a constant of proportionality, k.

$4 = k(12)^2$

$4 = 144k$

$k = \dfrac{4}{144} = \dfrac{1}{36}$

$a = \dfrac{1}{36} b^2$

When $b = 8$, $a = \dfrac{1}{36}(8)^2 = \dfrac{64}{36} = \dfrac{16}{9} = 1\dfrac{7}{9}$

NAILIT!

Always check back to see how the question asks the answer to be given, in this case a mixed number.

Currency problems

Converting between currencies is a particular example of direct proportion. An amount of money in one currency is proportional to the amount in another currency.

WORKIT!

Alegra is on holiday in the USA and sees a watch for \$120.

The exchange rate is £1 = \$1.45.

a Work out the cost of the watch in pounds.

Cost of watch in pounds $= \dfrac{120}{1.45} = £82.76$

Amount in \$ = 1.45 × amount in £.

> Commission is a charge that is a percentage of the value of the amount of money changed.

b When Alegra returns to the UK she has \$300 left over which she wants to change back into pounds.

The travel agent offers an exchange rate of £1 = \$1.32 without commission. The bank offers an exchange rate of £1 = \$1.28 with 3% commission.

Should she use the travel agent or the bank to convert her money back into pounds? Explain your answer.

Travel agent: Amount in pounds $= \frac{300}{1.32} = £227.27$

Bank:　　　　Amount in pounds $= \frac{300}{1.28} = £234.38$

> The 3% commission is worked out and then deducted.

Commission $= \frac{3}{100} \times 234.38 = £7.03$

Net amount from bank $= £234.38 - £7.03 = £227.35$

She should use the bank.

Inverse proportion

WORKIT!

It takes 3 men 5 hours to put up a fence. How long would it take 6 men?

> As the number of men doubles, the number of hours halves.

$5 \div 2 = 2.5$ hours

In the previous example, the number of men is **inversely proportional** to the time it takes. If y is **inversely proportional** to x, it can be written as $y \propto \frac{1}{x}$. This means if x doubles, the value of y halves. Inverse proportion can be written with a constant of proportionality, k, as $y = \frac{k}{x}$.

As for direct proportion, if you know a pair of values for x and y, then you can find the value of k.

WORKIT!

Here are some statements. For each statement, write down an equation including a constant.

a a is directly proportional to b.　　　　　　　　$a = kb$

b R is inversely proportional to I.　　Also, $I = \frac{k}{R}$　→　$R = \frac{k}{I}$

c y is directly proportional to the square of x.　　$y = kx^2$

d The time of a swing of a pendulum, T, in seconds is directly proportional to the square root of its length, l, in metres.　　$T = k\sqrt{l}$

e a is inversely proportional to the square root of b.　　$a = \frac{k}{\sqrt{b}}$

WORKIT!

The pressure, P, of a gas is inversely proportional to its volume, V.
When $V = 12$, $P = 5$.

a Find a formula for P in terms of V.

$$P \propto \frac{1}{V} \text{ so } P = \frac{k}{V}$$

$$5 = \frac{k}{12}, \text{ giving } k = 60$$

> Substitute $P = 5$ and $V = 12$ to find the value of k.

$$P = \frac{60}{V}$$

b Calculate the value of P when $V = 6$.

$$P = \frac{60}{V}$$

$$= \frac{60}{6}$$

$$= 10$$

DOIT!

Design and annotate a revision card for finding the constant of proportionality.

✓ CHECKIT!

1 L is inversely proportional to n.

Carlos says that if you double n, then L will double.

Explain why Carlos is wrong.

2 y is directly proportional to x.

a Write an equation showing the relationship between x and y.

b When $y = 8$, $x = 3$. Find the value of y when $x = 4$. Give your answer to 1 d.p.

3 Rosie is on holiday in Greece. She sees a pair of sunglasses costing €120.

She looks on the internet and finds that the cheapest price in the UK is £89.

The exchange rate is £1 = €1.27.

a In which country are the sunglasses cheaper?

b Work out the difference in the cost of the sunglasses in the UK and Greece. Give your answer in £.

4 The volume, $V\,\text{cm}^3$, of a sphere is directly proportional to the cube of its radius, $r\,\text{cm}$.

A sphere with a radius of 2 cm has a volume of $33.5\,\text{cm}^3$.

Find the volume of a sphere with a radius of 4 cm.

5 The pressure of a gas, P, is inversely proportional to the volume it occupies, V.

When $V = 1$, $P = 100\,000$. Work out P when $V = 3$. Give your answer to the nearest whole number.

6 $a \propto b^2$ and $a = 96$ when $b = 4$. Find a when $b = 5$.

7 The surface area of a cube, $A\,\text{cm}^2$, is directly proportional to the square of the side of the cube, $x\,\text{cm}$.

a Write the relationship as an equation and find the value of the constant of proportionality.

b Hence write down the surface area of a cube with side of length 4 cm.

Graphs of direct and inverse proportion and rates of change

Direct proportion

A graph showing direct proportion is a straight line passing through the origin.

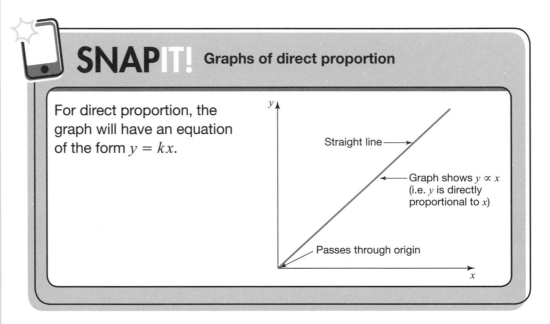

SNAP IT! Graphs of direct proportion

For direct proportion, the graph will have an equation of the form $y = kx$.

Straight line →

Graph shows $y \propto x$ (i.e. y is directly proportional to x)

Passes through origin

This equation is of the form $y = mx + c$, a straight line with gradient m and intercept on the y-axis c. The line passes through the origin, so $c = 0$. So the value of k will be the gradient of the line.

Inverse proportion

The graph showing inverse proportion is a curve of the type shown below.

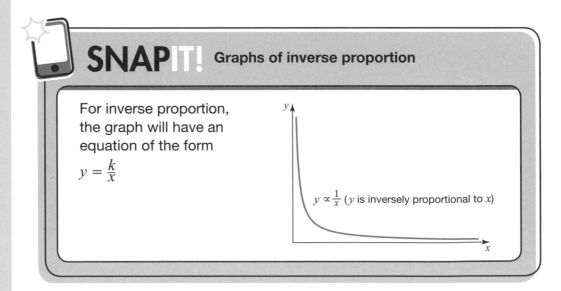

SNAP IT! Graphs of inverse proportion

For inverse proportion, the graph will have an equation of the form

$$y = \frac{k}{x}$$

$y \propto \frac{1}{x}$ (y is inversely proportional to x)

WORKIT!

A graph is drawn of a quantity V against a quantity c.

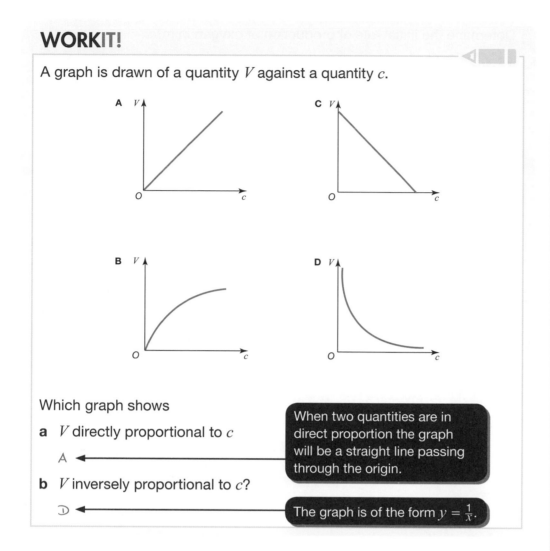

Which graph shows

a V directly proportional to c

A ◄————————

When two quantities are in direct proportion the graph will be a straight line passing through the origin.

b V inversely proportional to c?

D ◄————————

The graph is of the form $y = \frac{1}{x}$.

Calculating the rate of change from a graph

Graphs used to determine the **rate of change** of a quantity plotted on the y-axis always have **time** along the x-axis.

Suppose an experiment is conducted to investigate the rate of production of oxygen gas during a chemical reaction. The volume of oxygen in m³ is on the y-axis and the time in seconds is on the x-axis. The rate of change of the volume of oxygen is found by measuring the gradient of the graph.

WORKIT!

The volume of oxygen (O_2) produced in a chemical reaction was measured at certain times and gave the following results.

Time (s)	0	50	100	150	200	250	300
Volume O_2 (m³)	0	5.0	10.0	14.8	19.0	22.5	25.0

Plot these results on a graph with volume of oxygen on the y-axis and time on the x-axis.

Determine the initial rate of production of oxygen in m³/s.

> Make sure that you label both axes and include the units.

> The points are almost in a straight line initially but then they start to bend.
>
> Draw a straight line up to about 200 s and then a smooth curve after that. As the start of the graph is a straight line we can say that the gradient and hence the rate of change of volume of O_2 is constant.

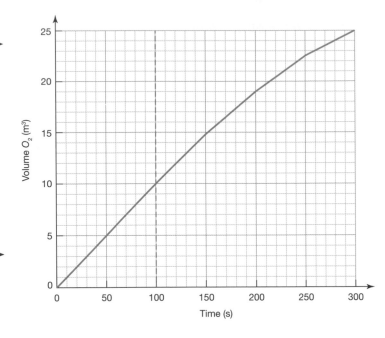

Initial rate of production of oxygen = gradient = $\frac{10}{100}$ = 0.1 m³/s

> To find the gradient, draw a triangle underneath the straight section of the graph.

DO IT!

Draw a poster of annotated graphs showing direct and inverse proportion.

CHECK IT!

1 Two quantities, c and V, are in inverse proportion. Which graph shows this?

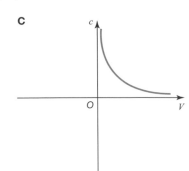

2 Quantity a is directly proportional to quantity b. Which one of the following graphs shows this?

A

B

C

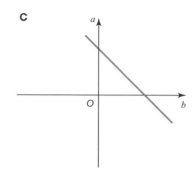

3 The graph shows two quantities, P and V, that are inversely proportional to each other.
The points $A(3, 12)$ and $B(6, a)$ lie on the curve.
Find the value of a.

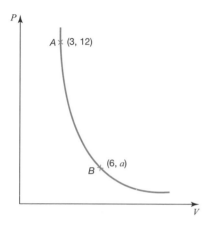

4 The graph shows two quantities x and y which are in inverse proportion to each other.
Find the values of coordinates a and b.

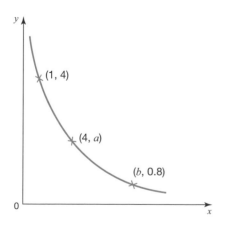

5 For the values below, y is directly proportional to x^2. If a is a positive number, work out the value of a.

x	2	a
y	16	36

Growth and decay

Compound interest

With **compound interest**, the interest is added onto the total amount at the end of each year, which then earns interest too. This means the overall increase is greater each year.

SNAP IT! Compound interest

Amount at the end of n years = $A_0 \times$ (multiplier)^n
where n = time period in years, A_0 = original amount of money,
and multiplier = $1 + \frac{\% \text{ interest rate}}{100}$.

Use the formula with $A_0 = 2000$ and $n = 3$.

As this is a money question, give your answer to the nearest penny.

WORKIT!

Jack invests £2000 at 2.75% compound interest. Find the value of the investment at the end of 3 years.

Multiplier = $1 + \frac{2.75}{100} = 1.0275$ ← Work out the multiplier.

Amount at the end of 3 years = $2000 \times (1.0275)^3 = £2169.58$

Exponential growth

Compound interest is one example of **exponential growth**. Other examples can be found in biology. The formula is the same, although the periods of time may be days or hours, not years.

SNAP IT! Exponential growth

Amount at the end of n periods of time = $A_0 \times$ (multiplier)^n
where n = number of time periods (e.g. years),
A_0 = original amount, and multiplier = $1 + \frac{\% \text{ rate of growth}}{100}$

WORKIT!

1 A colony of bacteria increases by 12% each day. There are 4000 bacteria at the start. How many bacteria will there be after 3 days? Give your answer to the nearest whole number.

Multiplier = $1 + \frac{12}{100} = 1.12$

Amount at the end of 3 days = $4000 \times 1.12^3 = 5620$ (to 4 s.f.)

2 Algae in a pond grow very quickly in warm weather. The number of
algae is 200 000 after 5 days. If the algae grow at a rate of 23% per
day, how many algae were there originally?

Multiplier $= 1 + \frac{23}{100} = 1.23$ $200\,000 = A_0 \times (1.23)^5$

Amount at the end of n days $\frac{200\,000}{(1.23)^5} = A_0$
$= A_0 \times$ (multiplier)^n

$A_0 = 71\,040$

> Substitute the amount
> after 5 days and $n = 5$.

Exponential decay

Exponential decay is where a quantity goes down by a
certain percentage – like exponential growth but in reverse.
For example, when you buy a new car it depreciates (i.e.
the value goes down with time).

For exponential decay, the formula is the same as for
exponential growth but the multiplier is less than 1.

SNAPIT!

Exponential decay

Amount at the end of n years
$= A_0 \times$ (multiplier)^n

where $n =$ time period in years,
$A_0 =$ original amount and

multiplier $= 1 - \dfrac{\text{% rate of decay}}{100}$

WORKIT!

Natalie buys a computer for £850. Its value depreciates by 30% in
the first year and 20% in each of the next 2 years. Find the value of
the computer after 3 years, giving your answer to the nearest whole
number.

Multiplier for year 1 $= 1 - \frac{30}{100} = 0.7$

Multiplier for years 2 and 3 $= 1 - \frac{20}{100} = 0.8$

Value after 3 years $= 850 \times 0.7 \times 0.8 \times 0.8 = £380.80$

> For decay questions,
> the multiplier is
> calculated by
> **subtracting** the rate
> from 1.

NAILIT!

The percentage
change may vary over
time. Then you need
to use the multiplier
for each period.

CHECKIT!

1 Find the multiplier for each of these
situations.

a pond weed in a pond where the
growth rate is 5% per week

b bacterial growth where growth rate is
25% per hour

c increase in money if the rate of
compound interest is 3.75% per year

d decrease in value if the rate of
depreciation is 21%.

2 A motorbike depreciates at a rate of 18%
per year. Ben buys a new motorbike
costing £9000. How much will his

motorbike be worth after 3 years?
Give your answer to the nearest whole
number.

3 Amir put £650 in a savings account.
The rate for the first 2 years was 4%
compound interest and for the next
2 years was 3% compound interest. How
much did he have in the account after
4 years?

4 A restaurant chain currently has 4000
restaurants across the UK. It has been
expanding by 25% each year. How
many restaurants did the company
have 3 years ago?

Ratios of lengths, areas and volumes

Scale factors

When two shapes are **similar**, they are identical in shape but not in size. One shape is simply an **enlargement** of the other. A missing length on one of the shapes can be found by considering the scale factor:

Scale factor for enlargement $= \dfrac{\text{big}}{\text{small}}$

Scale factor for reduction $= \dfrac{\text{small}}{\text{big}}$

WORKIT!

P and Q are similar triangles. Find the length of the side marked x on triangle Q.

You are finding the smaller length, so find the scale factor for the reduction.

$$\text{Scale factor} = \frac{\text{corresponding side of Q}}{\text{corresponding side of P}} = \frac{8}{12} = \frac{2}{3}$$

$$x = \frac{2}{3} \times 9 = 6 \, cm$$

Multiply the side on P corresponding to x by the scale factor.

You can prove two triangles are similar by comparing the ratios of corresponding sides.

WORKIT!

Identify whether these triangles are similar.

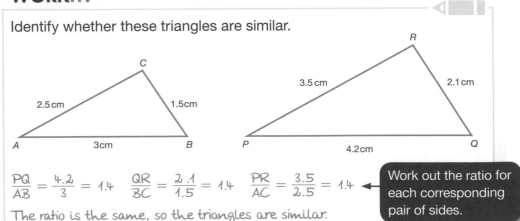

$$\frac{PQ}{AB} = \frac{4.2}{3} = 1.4 \quad \frac{QR}{BC} = \frac{2.1}{1.5} = 1.4 \quad \frac{PR}{AC} = \frac{3.5}{2.5} = 1.4$$

Work out the ratio for each corresponding pair of sides.

The ratio is the same, so the triangles are similar.

Scale factors for area and volume

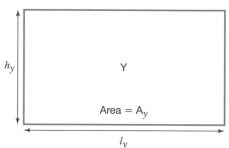

For rectangles X and Y, scale factor for length, $k = \dfrac{h_y}{h_x} = \dfrac{l_y}{l_x}$

Scale factor for area $= \dfrac{h_y \times l_y}{h_x \times l_x}$

$= \dfrac{h_y}{h_x} \times \dfrac{l_y}{l_x} = k^2$ or the square of the scale factor for length.

Similarly, the scale factor for volume is equal to the cube of the scale factor for length.

SNAPIT!

2D and 3D scale factors

For two similar shapes or solids with a scale factor k:

- scale factor for area $= k^2$

- scale factor for volume $= k^3$

WORKIT!

Prove that triangle *ABC* is mathematically similar to triangle *DBE*.

$AD = 3\,cm$, $DB = 6\,cm$ and $AC = 8\,cm$.

Work out the area of triangle *DBE*.

Angle CAB = angle EDB = 90°

Angle ABC = angle DBE (same angle)

Angle ACB = 180° − angle CAB − angle ABC
(angle sum in a triangle)

Angle DEB = 180° − angle EDB − angle DBE (angle sum in a triangle)

So angle ACB = angle DEB

All the angles are the same, so triangles ABC and DBE are similar.   ◄── First prove that the triangles are similar.

Area of triangle $ABC = \dfrac{1}{2} \times b \times h = \dfrac{1}{2} \times 9 \times 8 = 36\,cm^2$

Scale factor $= \dfrac{DB}{AB} = \dfrac{6}{9} = \dfrac{2}{3}$ ◄── Work out the scale factor from the ratio of corresponding sides.

$\dfrac{Area\ DBE}{Area\ ABC} = (scale\ factor)^2 = \left(\dfrac{2}{3}\right)^2 = \dfrac{4}{9}$

Area $DBE = \dfrac{4}{9} \times Area\ ABC = \dfrac{4}{9} \times \dfrac{1}{2} \times 9 \times 8 = 16\,cm^2$ ◄── Use the fact that the triangles are similar to write the ratio of the areas.

NAILIT!

If you are not told two triangles are similar, you must first prove that they are similar.

WORKIT!

Two solid spheres, P and Q, are mathematically similar.
The ratio of the volume of P to the volume of Q is 8 : 125.

a Find the ratio of the radius of sphere P to the radius of sphere Q.

$$\frac{V_P}{V_Q} = \left(\frac{r_P}{r_Q}\right)^3$$

> Scale factor for volume = (scale factor for length)³.

$$\frac{8}{125} = \left(\frac{r_P}{r_Q}\right)^3$$

$$\frac{r_P}{r_Q} = \sqrt[3]{\frac{8}{125}} = \frac{2}{5}$$

b The surface area of sphere Q is 37.5 cm². Show that the surface area of sphere P is 6 cm².

> Scale factor for area = (scale factor for length)².

$$\frac{A_P}{A_Q} = \left(\frac{r_P}{r_Q}\right)^2 = \left(\frac{2}{5}\right)^2 = \frac{4}{25}$$

$$A_P = \frac{4}{25} \times A_Q = \frac{4}{25} \times 37.5 = \frac{150}{25} = 6$$

Surface area of sphere P is 6 cm².

DOIT!

Make up some scale factor questions then come back to them tomorrow.

CHECKIT!

1 Triangles A and B are similar. Find the length of the side marked x on triangle A. Give your answer as a mixed number.

A
x
6cm

B
10 cm
9 cm

2 Two cylinders are mathematically similar. Cylinder A has a volume of 27 cm³ and cylinder B has a volume of 64 cm³.

The surface area of cylinder B is 96 cm².

Show that the surface area of cylinder A is 54 cm².

3

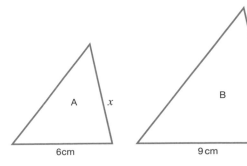

ABC and AED are straight lines. Lines BE and CD are parallel. Angle $ACD = 90°$.

$CD = 8$ cm, $BC = 5$ cm and $AB = 5$ cm.

a Work out the length of side BE.

b Work out the area of triangle ABE.

Gradient of a curve and rate of change

The gradient of a curve changes with position so you need to know where the gradient is to be measured. Usually the x coordinate is given, which determines where on the curve to find the gradient.

With rate of change graphs, time is always plotted on the x-axis.

The instantaneous rate of change

The **instantaneous rate of change** is the gradient of the tangent drawn at that point.

For example, this graph shows how the velocity of an object changes with time.

The gradient of a velocity–time graph gives the acceleration. To find the acceleration at time $t = 10\,\text{s}$, work out the gradient of the tangent:

Acceleration at $10\,\text{s}$ = gradient
$= \dfrac{12 - 4}{16 - 6} = \dfrac{8}{10} = 0.8\,\text{m/s}^2$

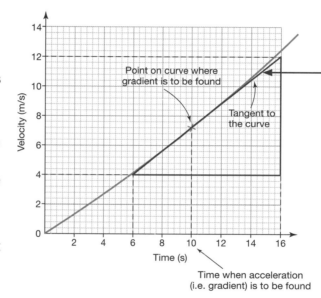

Point on curve where gradient is to be found

Tangent to the curve

Time when acceleration (i.e. gradient) is to be found

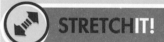

The tangent is the straight line that just touches the curve.

STRETCH**IT!**

Use the same method to find the gradient of a curve from a graph by drawing in a tangent.

The average rate of change

The **average rate of change** is the gradient of the line drawn between two points on the curve. Each point corresponds to a certain time.

Average acceleration between $6\,\text{s}$ and $16\,\text{s}$ = gradient = $\dfrac{8}{10}$
$= 0.8\,\text{m/s}^2$

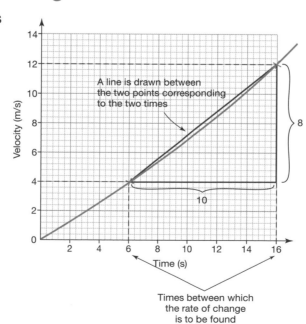

A line is drawn between the two points corresponding to the two times

Times between which the rate of change is to be found

WORKIT!

The graph shows how the height of grain in a large hopper changes with time, t, when it is being filled with grain.

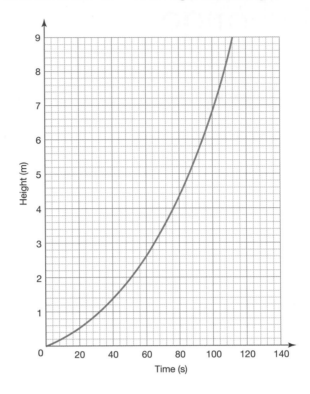

a Find the instantaneous rate of change of height when $t = 80$ s.

Draw a tangent to the curve at $t = 80$ s and find its gradient.

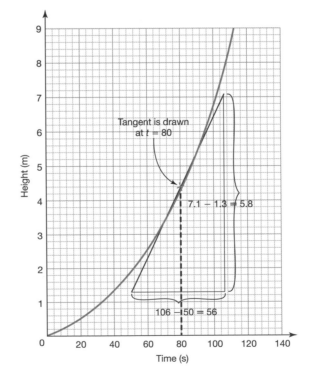

Instantaneous rate of change of height = gradient = $\dfrac{5.8}{56}$

= 0.1 m/s (to 1 s.f.)

b Find the average rate of change of height between $t = 40\,\text{s}$ and $t = 80\,\text{s}$.

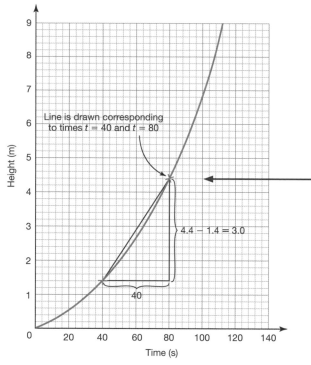

Line is drawn corresponding to times $t = 40$ and $t = 80$

$4.4 - 1.4 = 3.0$

40

Join the points on the curve at $t = 40\,\text{s}$ and $t = 80\,\text{s}$ with a straight line and find its gradient.

Average rate of change of height between $t = 40\,\text{s}$ and $t = 80\,\text{s}$

$\quad = \text{gradient} = \dfrac{3}{40} = 0.08\,\text{m/s}$ (to 1 s.f.)

DOIT!

Draw and annotate distance–time and velocity–time graphs to show instantaneous and average gradients and what they represent.

CHECKIT!

1 An accelerating particle has the velocity–time graph shown on the right.

a State what the gradient of the curve represents.

b Find the average acceleration of the particle between 2 s and 6 s.

c Find the time when the instantaneous acceleration is the same as the average acceleration calculated in part b.

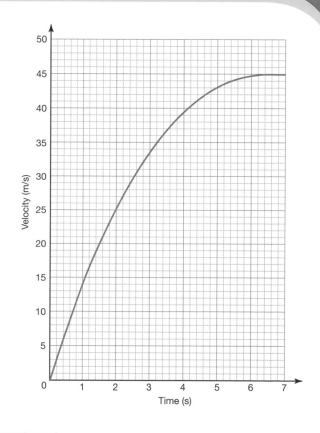

Converting units of areas and volumes, and compound units

Converting areas and volumes

Converting between m² and cm²

A 1 m by 1 m square has an area of 1 m².

This is the same as a 100 cm by 100 cm square, which has an area of 100 × 100 = 10 000 cm².

> **Note that 1 m² does not equal 100 cm².**

> **We've given one answer in standard form and one as a decimal. Both are acceptable.**

So 1 m² = 10 000 cm² and 1 cm² = $\frac{1}{10\,000}$ m² = 1 × 10⁻⁴ m².

Similarly, 1 cm² = 100 mm² and 1 mm² = $\frac{1}{100}$ cm² = 0.01 cm².

Converting between m³ and cm³

Similarly, a cube of side 1 m has a volume of 1 m³, and a cube of side 100 cm has a volume of 100 × 100 × 100 = 1 000 000 cm³.

So 1 m³ = 1 000 000 cm³ and 1 cm³ = $\frac{1}{1\,000\,000}$ m³ = 1 × 10⁻⁶ m³.

Similarly, 1 cm³ = 1000 mm³ and 1 mm³ = $\frac{1}{1000}$ cm³ = 0.001 cm³.

WORKIT!

1 Convert:

   a 250 cm³ to m³

      250 cm³ = 250 × 10⁻⁶ m³ = 2.5 × 10⁻⁴ m³

   b 1.500 m² to cm²

      1.5 m² = 1.5 × 10 000 cm² = 1.5 × 10⁴ cm²

> **The per symbol (i.e. /) acts like a divide sign, and can tell you information about the formula. For example, one unit of speed is km/h, which tells you a distance in km is divided by a time in h.**

Compound units

Compound units are made up of two units (for example m/s, km/h, g/cm³). Quantities such as speed, density, pressure, rate of pay and price per unit have compound units. These are all worked out by dividing one quantity with one unit by a different quantity with a different unit.

SNAPIT! Compound units

Compound units include:

$$speed = \frac{distance}{time}$$ unit m/s or km/h

$$density = \frac{mass}{volume}$$ unit g/cm³ or kg/m³

$$pressure = \frac{force}{area}$$ unit N/m² or Pa (= N/m²)

$$rate\ of\ pay = \frac{pay}{time}$$ unit £/h, £/month, £/year

$$price\ per\ unit = \frac{price}{number\ of\ units}$$ unit for price on the top could be pence or pounds
unit on the bottom could be g, kg, ml, tonne, etc.
compound unit p/g, £/kg, etc.

Speed is worked out using the formula

You need to be able to rearrange this

formula to get: distance = speed × time

$$speed = \frac{distance}{time}$$

$$time = \frac{distance}{speed}$$

NAILIT!

You need to learn this formula as it will not be given. To remember the formula for speed, think of the units for speed, which are m/s or km/h: distance divided by time.

STRETCHIT!

When speed is in a particular direction, it is called **velocity**. For information on velocity-time graphs see pages 92–3.

WORKIT!

1 Dhaya is driving her car at an average speed of 80 km/h.

a Work out her speed in metres per second, correct to the nearest whole number.

$80\,km/h = 80 \times 1000\,m/h$

$\qquad = \frac{80 \times 1000}{60 \times 60}\,m/s$

$\qquad = 22.22\,m/s$

$\qquad = 22\,m/s$ (to nearest whole number)

NAILIT!

Always look back at the question to check how accurate the answer needs to be.

b Calculate the time in minutes it would take her to drive a distance of 30 km at this average speed.

$time = \frac{distance}{speed} = \frac{30}{80} = 0.375\,h$ ◀

Rearrange $speed = \frac{distance}{time}$ to give $time = \frac{distance}{speed}$

$0.375\,h = 0.375 \times 60\,minutes = 22.5\,minutes$

NAILIT!

Always look at the units required for the answer. This tells you whether you need to change any of the units for the values in the calculation.

The answer needs to be in N/m², so convert the area from cm² to m².

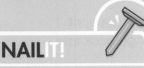

Remember that time is not metric.

2 Using the formula pressure = $\frac{\text{force}}{\text{area}}$ work out the pressure in N/m² when a force of 500 N acts on an area of 20 cm².

$$20\,cm^2 = 20 \times 10^{-4}\,m^2 = 2 \times 10^{-3}\,m^2$$

$$pressure = \frac{force}{area} = \frac{500}{2 \times 10^{-3}} = 250\,000\,N/m^2$$

3 A train covers a distance of 29 km in 45 minutes. Calculate its average speed in km/h. Give your answer to the nearest whole number.

$$45\,mins = \frac{45}{60}\,h = 0.75\,h$$

$$speed = \frac{distance}{time} = \frac{29}{0.75} = 38.67\,km/h = 39\,km/h \text{ (to nearest hour)}$$

DO**IT!**

Look up the 'speed formula triangle' on the internet. Produce similar triangles for density and pressure.

✓ CHECK**IT!**

1 The diagram shows a solid block of wood.

4 cm   5 cm   6 cm

a Find the total surface area of the block in:

i mm²   **ii** m².

b Find the volume of the block in m³.

2 A bar of gold has a volume of 600 cm³ and a mass of 1.159 kg. Work out the density of gold in g/cm³ giving your answer correct to 3 decimal places.

3 A block of steel with a volume of 1 m³ is melted and used to create lots of ball bearings. The volume of each ball bearing is 0.5 cm³. How many ball bearings could be made?

4 A car is travelling with an average speed of 50 km/h. Find this speed in m/s. Give your answer correct to 2 decimal places.

5 Kieran drove from Manchester to Bristol. It took him 4 hours at an average speed of 70 km/h.

Mary drove by the same route and took 5 hours.

a What was Mary's average speed for her journey?

b If Mary took a different route, how might this affect your answer to part a?

1 In a sale the price of a dress was reduced by 20% to £76.80. What was the original price of the dress?

2 The area of floating pond weed that grows on the surface of a pond in t days is $A\,\text{m}^2$, where

$A_0 = 5$

$A_{t+1} = 1.02A_t$

Work out the area covered by the pond plants in

a 1 day   **b** 3 days.

Give your answers to 2 decimal places.

3 y is inversely proportional to x.

When $x = 2.5$, $y = 4$.

Find the value of y when $x = 5$.

4 The ratio of the number of students studying history in Year 11 to those not studying history is $2:7$. There are 54 students studying history. How many students are there in Year 11?

5 The price of a house in a certain town has increased by 120% over the last 10 years.

The original price of the house was £220 000. How much is the house worth now? Give your answer to the nearest £5000.

6 £2000 is invested at an interest rate of 2.5% for 5 years. Work out how much money is in the account after 5 years if the interest paid is

a simple interest

b compound interest (to the nearest penny).

7 Tom goes on holiday to China. The exchange rate is £1 = 8.55 yuan.

He changes £400 into yuan.

a Work out how many yuan he should get.

After his holiday, Tom has 800 yuan left over which he wants to change back into pounds.

The travel agent offers an exchange rate of £1 = 8.6 yuan with 2.5% commission.

The post office offers an exchange rate of £1 = 8.9 yuan with no commission.

b Should Tom change his money at the travel agent or the post office? Explain your answer.

8 A company is owned by Joshua, Amy and Luke, in the ratio of $3:5:7$.

Any profits are shared in this ratio.

When the profits in a certain year were shared, Luke received £4000 more than Amy.

Work out the total profit for the year.

9 Two solids are mathematically similar. Solid A has a surface area of $25\,\text{cm}^2$ and solid B has a surface area of $4\,\text{cm}^2$.

The volume of solid A is $10\,\text{cm}^3$.

Find the volume of solid B.

10 The number of male and female members of a swimming club is in the ratio $7:4$.

25% of the male members are junior members.

10% of the female members are junior members.

Work out the percentage of junior members in the club, giving your answer to the nearest integer.

11 A bronze alloy is made by mixing copper and tin. The masses of the copper and tin are in the ratio $9:1$.

The density of copper is $8.9\,\text{g/cm}^3$.

The density of tin is $7.3\,\text{g/cm}^3$.

Work out the density of the bronze alloy. Give your answer to 1 decimal place.

Geometry and measures

2D shapes

You need to be familiar with the following terms.

Polygon: a figure with straight edges.

Irregular polygon: a polygon where the sides and angles are not all equal. For example, the diagram shows an irregular hexagon.

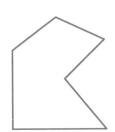

Regular polygon: a polygon where all the sides and angles are equal.

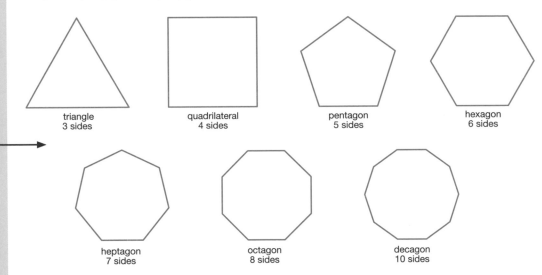

triangle
3 sides

quadrilateral
4 sides

pentagon
5 sides

hexagon
6 sides

heptagon
7 sides

octagon
8 sides

decagon
10 sides

> 'You may also come across a nonagon (9 sides), a decagon (10 sides), or a dodecagon (12 sides).

> For regular polygons, the order of rotational symmetry and the number of lines of symmetry are both the same as the number of sides.

Order of rotational symmetry: the number of times you can rotate the shape onto itself so it fits exactly in one complete revolution.

Lines of symmetry: the number of mirror lines.

> These two are also **scalene** triangles if their angles and sides are all different.

SNAP IT! Types of triangle

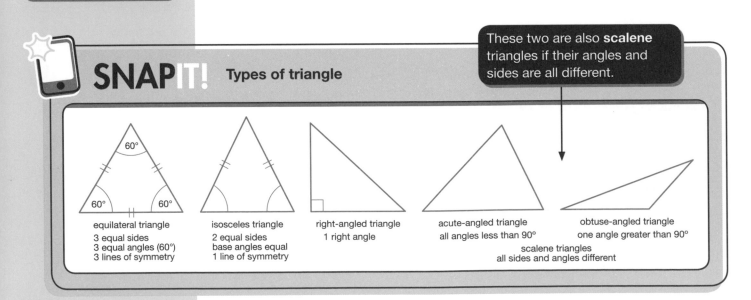

equilateral triangle
3 equal sides
3 equal angles (60°)
3 lines of symmetry

isosceles triangle
2 equal sides
base angles equal
1 line of symmetry

right-angled triangle
1 right angle

acute-angled triangle
all angles less than 90°

obtuse-angled triangle
one angle greater than 90°

scalene triangles
all sides and angles different

SNAP**IT!**   Types of quadrilateral

square

rectangle

rhombus

parallelogram

kite

trapezium

trapezium

Name	Pairs of parallel sides	Lines of symmetry	Order of rotational symmetry	Diagonals same length?	Equal sides	Equal angles
Square	2	4	4	Yes	Yes	All 90°
Rectangle	2	2	2	Yes	2 pairs	All 90°
Rhombus	2	2	2	No	Yes	2 pairs
Parallelogram	2	0	2	No	2 pairs	2 pairs
Kite	0	1	1	No	2 pairs	1 pair
Trapezium	1	1 or 0	1	Can be	1 pair or none	2 pairs, 1 pair or none

With a square, rhombus or kite, the diagonals intersect at right angles. For a square they are the same length, but for the other two shapes they are not.

DO**IT!**

Draw a flowchart for finding out the type of triangle or quadrilateral from information about its sides, angles and symmetry.

CHECK**IT!**

1 Here are some statements about polygons. State whether each statement is true or false.

   a A trapezium always has one pair of parallel sides.

   b The diagonals of a rhombus are the same length.

   c It is possible to have a right-angled isosceles triangle.

   d A parallelogram has order of rotational symmetry of 1.

   e The order of rotational symmetry of a regular polygon is the same as the number of sides.

   f An equilateral triangle can only have internal angles of 60°.

Constructions and loci

Constructions

Using only a ruler and compasses, you will be asked to **construct** angles, lines and figures. Here are the constructions you need to be able to draw.

Perpendicular bisector of a line segment

To bisect (cut in half) a line with another line at right angles:

A bisector divides something into two equal parts.

1. Put the compass point on A and set the radius to about two-thirds of AB.

2. Without altering the radius, draw two arcs at approximate positions P and Q.

3. Repeat step 2, putting the compass point on B.

4. Draw the line between P and Q.

   The line PQ is the set of points equidistant (equal in distance) from the points A and B.

Perpendicular to a line at a given point

To construct a line passing through point P and perpendicular to AB:

1. Put the compass point on P and set the radius to draw two arcs cutting AB.

2. Alter the radius so that it is about half the distance from point P to the line.

3. Using this radius, draw an arc from each of the intersections of the arcs from step 1.

4. Draw a line from point P through the intersection of the arcs until it meets AB.

The perpendicular from P is the shortest distance to the line AB.

Perpendicular to a line through a given point on the line

To draw the perpendicular through a point on a line, use the same radius to draw two arcs on the line centred on P. Then use these two intersections as A and B and construct the perpendicular bisector of AB.

Constructing angles

Angle of 60°

To construct an angle of 60°:

1. Draw a line through the point P where you want to construct the angle (if there isn't one already drawn).

2 Put the compass point on *P* and draw an arc through about a quarter circle that finishes at the line.

3 Without altering the radius, draw an arc from the point where the arc in step 1 intersects the line.

4 Draw a line from P through the intersection of the arcs. The angle between the two lines is 60°.

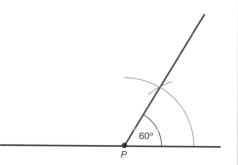

Bisecting an angle

To bisect (cut in half) angle *AOB*:

1 Put the compass point on *O* and draw two arcs cutting *OA* and *OB*.

2 Without altering the radius, draw two arcs from the intersections in step 1, to intersect at *P*.

3 Draw a line through *P* to *O*.

> The angle bisector is the set of points equal in distance from both lines.

 STRETCHIT!

To construct an angle of 30°:
1 Construct an angle of 60°.
2 Bisect this angle.

To construct an angle of 45°:
1 Construct an angle of 90°.
2 Bisect this angle.

Loci

If the position of a point *P* moves in some way then all the possible positions point *P* can take is called the **locus** (plural **loci**) of *P*.

WORKIT!

AB is a straight line. Draw the locus of the point *P* that moves in such a way that it is always 2 cm from *AB*.

> **NAILIT!**
>
> A locus that stays the same distance from a point is either a circle or an arc of a circle.

> Draw two semicircles of radius 2 cm with centres *A* and *B*. Join the tangents to these to give the locus.

WORKIT!

Use only ruler and compasses to answer this question.

Point *P* lies inside the quadrilateral *ABCD*. It is the same distance from *A* and *B* and also a distance of 7 cm from point *D*. Mark on the diagram the position of *P*.

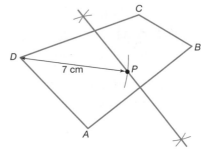

Draw the perpendicular bisector of *AB* to give all the points equal distance from *A* and *B*.

Draw an arc centred on *D* with radius 7 cm to give all the points 7 cm from *D*.

P is where the line and arc intersect.

DOIT!

Write out a revision card for each construction.

WORKIT!

Two supermarkets *A* and *B* are 20 km apart. Supermarket *A* delivers to houses up to 12 km away. Supermarket *B* delivers to houses that are within 10 km.

Draw a scale diagram to show the area where both supermarkets could deliver. Show the area by shading it.

Use the scale of 1 cm to 4 km.

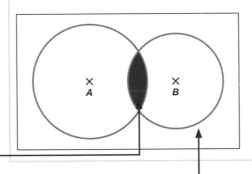

If the radius of the larger circle is a and the radius of the smaller circle is b:

$$\frac{a}{1} = \frac{12}{4} = 3 \text{ cm}$$

$$\frac{b}{1} = \frac{10}{4} = 2.5 \text{ cm}$$

Shade the region that lies within both circles.

This drawing is not to scale, but in your drawing the larger circle (centre *A*) should have a radius of 3 cm and the smaller circle (centre *B*) should have a radius of 2.5 cm. *A* and *B* should be 5 cm apart.

CHECKIT!

1 *OP* and *OQ* are straight lines at an angle to each other. *X* travels in such a way that it is always the same distance between *OP* and *OQ*.

Copy the diagram and draw the locus of *X*.

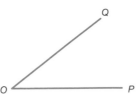

2 Two towns, *A* and *B*, are 70 km apart.

A wind farm is to be sited in an area around the two towns. It must be at least 30 km from town *A* and at least 60 km from town *B*.

Using a scale of 1 cm to 10 km, draw a map and shade the region where the wind farm could be situated.

Properties of angles

SNAP IT! Angle properties

There are lots of angle properties to remember. Take a picture of each one to help you revise.

Angles at a point

Angles at a point on a straight line add up to 180°.
$x + y = 180°$

Vertically opposite angles are equal.

Angles at a point add up to 360°.
$a + b + c = 360°$

NAIL IT!

Remember the angles of a triangle always add up to 180° and so do the angles at a point on a straight line.

Properties of triangles

The angles in a triangle add up to 180°.
$a + b + c = 180°$

Isosceles triangle: equal base angles. The top angle can be found by adding the two base angles together and subtracting the result from 180°.

Equilateral triangle: all angles are 60° and all sides are equal.

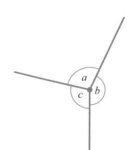

Parallel lines

Parallel lines are the same distance apart along their whole length.

Alternate angles are equal; on opposite sides of the line cutting the parallel lines; sometimes called 'Z' angles.

Corresponding angles are equal; sometimes called 'F' angles.

NAIL IT!

As soon as you see parallel lines, look for alternate or corresponding angles.

WORKIT!

AB and *CD* are two parallel lines.

a Work out, stating reasons, the size of angle *STR*.

angle *STR* = angle *BRT* = 70° (alternate angles)

> Angle *STP* is the same as angle *STR*.

b Work out, stating reasons, the size of angle *x*.

$x = 180 -$ angle *PST* $-$ angle *STP* (angle sum of a triangle)

$= 180 - 50 - 70 = 60°$ (*STP* is the same as *STR*)

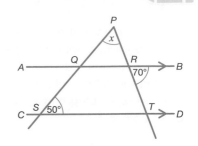

NAIL IT!

To prove that angles in a quadrilateral add up to 360°, draw a diagonal to divide a quadrilateral into 2 triangles.

Proving that the angles in a triangle add up to 180°

You need to be able to prove that the angle sum of a triangle is 180°.

The angles marked *a* are alternate angles.

The angles marked *b* are alternate angles.

c is the top angle of the triangle.

$a + b + c = 180°$ (angles on a straight line)

This is also the total of the angles in the triangle, so the angles in a triangle add up to 180°.

> Draw two parallel lines as shown.

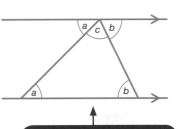

Angles in regular polygons

> Remember that regular polygons have equal sides and angles.

SNAP IT!   Angles in regular polygons

$$\text{Exterior angle} = \frac{360°}{\text{number of sides}}$$

Interior angle + exterior angle = 180°

Total of all the interior angles = number of sides × interior angle

interior angle

exterior angle

sides of a regular polygon

> You need to spot that you need to find the interior angles of each of the polygons.

WORKIT!

The diagram shows a regular pentagon and a regular hexagon.

a Work out the size of the interior angle of a regular pentagon.

Exterior angle $= \frac{360°}{\text{number of sides}} = \frac{360}{5} = 72°$

Interior angle $= 180 - 72 = 108°$

b Hence work out the size of angle *x*. Show all your working.

Exterior angle of regular hexagon $= \frac{360°}{\text{number of sides}} = \frac{360}{6} = 60°$

Interior angle $= 180 - 60 = 120°$

Angle $x = 360 - (108 + 120) = 132°$

> The angles at a point all add up to 360°.

DO**IT!**

Make two copies of the diagram. On one diagram, colour each pair of alternate angles a different colour. On the other, colour each pair of corresponding angles a different colour.

CHECK**IT!**

1 Calculate the interior angle of a regular polygon with ten sides.

2 The exterior angle of a regular polygon is 40°.

  a Find the number of sides of the polygon.

  b Calculate the sum of all the interior angles of the polygon.

3 In the diagram, *AC* and *BE* are parallel. Find the size of angle *x*.

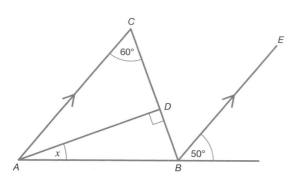

4 In the diagram, *AC*, *DF* and *GJ* are three parallel lines.

  a Write down, stating a reason, the size of angle *x*.

  b Write down, stating a reason, the size of angle *DEH*.

5 *ABCD* is a rhombus. Work out the angles of the rhombus.

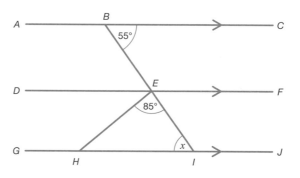

Congruent triangles

Two triangles are **congruent** if they fit exactly over each other. All the corresponding sides and corresponding angles are equal.

Triangles are congruent if at least three special facts are known.

SNAPIT!   Congruent triangles

The four possible **conditions for congruence** are that the following need to match:

- three sides (SSS)

- two sides and the angle between them (SAS)

- two angles and a corresponding side (ASA)

- for right-angled triangles only: right angle, hypotenuse and one other side (RHS)

Once you have established the congruence using one of the above cases, you can use the fact that all the corresponding sides and angles are equal to find unknown angles and sides.

WORKIT!

Here are some pairs of triangles. Say if each pair is congruent or not. If they are congruent, give the reason (e.g. SAS).

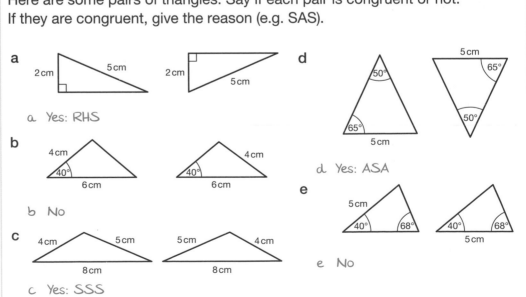

a

a Yes: RHS

b

b No

c

c Yes: SSS

d

d Yes: ASA

e

e No

WORKIT!

In the diagram, *ABCD* is a rhombus and *AE* = *FC*.

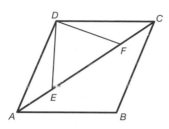

> Mark any equal sides and angles on a copy of the diagram.

Prove that triangle *AED* is congruent to triangle *CFD*.

AE = *FC* (given in the question)

AD = *CD* (sides of a rhombus equal)

So triangle *ACD* is isosceles, so the base angles *EAD* and *FCD* are equal.

Triangle *AED* is congruent to triangle *CFD* (SAS).

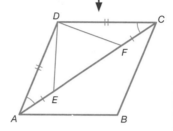

DOIT!

Use congruent triangles to prove two of the constructions you covered on pages 130–1.

DOIT!

Create a mnemonic or a song to help you remember the conditions for congruence.

NAILIT!

In questions involving rhombuses and parallelograms, look for isosceles triangles, corresponding angles and alternate angles to prove congruency.

CHECKIT!

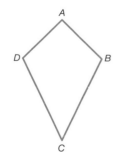

1 In the quadrilateral, *AB* = *AD* and *BC* = *CD*. Prove that angle *ABC* = angle *ADC*.

2 *ABC* is a triangle with *AB* = *AC*. *M* is the midpoint of *BC*.

Prove that angle *AMB* is a right angle.

Transformations

When a transformation is applied to a shape, the original shape is called the **object** and the shape after the transformation is called the **image**.

Translations

With a **translation**, the shape moves according to a **vector** $\begin{pmatrix} x \\ y \end{pmatrix}$. Each x coordinate of the shape moves x units to the right if x is positive, or x units to the left if x is negative. Each y coordinate of the shape moves y units up if y is positive, or y units down if y is negative.

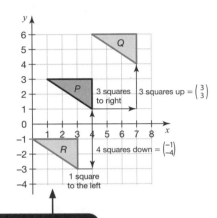

The diagram shows two different translations for the triangle P.

> Note that the shapes are identical under a translation (i.e. they are congruent).

Reflections

When a shape undergoes a **reflection** you need to know where the **mirror line** is.

Examples of mirror lines include:

- the y-axis (the vertical axis with the equation $x = 0$)

- the x-axis (the horizontal axis with the equation $y = 0$)

- $x = 4$ (a vertical line parallel to the y-axis at $x = 4$)

- $y = -2$ (a horizontal line parallel to the x-axis at $y = -2$)

- $y = x$ (a sloped line with a positive gradient, sloping at 45° if both axes have the same scale)

- $y = -x$ (a sloped line with a negative gradient, sloping at 45° if both axes have the same scale).

> Reflected shape after reflection in line $x = -2$.

> Reflected shape after reflection in line $y = x$.

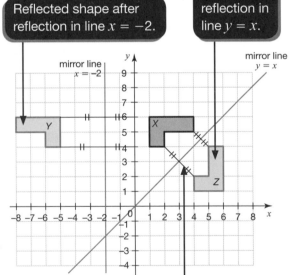

> Each point on the original shape is the same distance from the mirror line as the corresponding point on the image.

The diagram shows two reflections of shape X; one reflected in the line $x = -2$ and the other reflected in the line $y = x$.

Rotations

To describe a **rotation**, you need:

- the coordinates of the **centre** of rotation

- the **angle** of rotation (e.g. 90°, 180°, etc.)

- the **direction** of the rotation (clockwise or anticlockwise).

To find these look at the following diagram, where shape A is transformed to shape B. First you have to establish that it is a rotation. Any transformation that is not a reflection or translation must be a rotation.

1 Use a piece of tracing paper and trace shape A.

2 Put a pencil or pen point near to where you think the centre of rotation is.

3 Rotate the tracing paper to see if you can match up the triangle you drew with shape B. If you can't, try another point until the shape matches up.

4 Record the angle you moved the tracing paper from its original position, the coordinates of the centre you used and the direction (clockwise or anticlockwise).

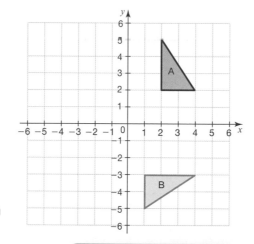

Shape A to shape B represents a rotation of 90° clockwise about the point (−1, 0).

NAIL IT!

Never assume that the origin is the centre of rotation as often it is not.

Enlargements

To describe an **enlargement**, you need:

The scale factor can be an integer or a fraction, and positive or negative.

- the coordinates of the centre of enlargement

- the scale factor. ◀——

The scale factor means how much all the lengths have been multiplied by.

To draw the enlargement of shape A, centre (0, 0) and scale factor 2:

1 Draw lines from the centre of enlargement through the corners of the shape.

A **vertex** is a corner. The plural of vertex is vertices.

2 Choose one vertex of shape A and ◀—— count how many squares along and up from the centre of enlargement: bottom left vertex is 3 right and 2 up.

3 Multiply the distance across and up or down by the scale factor. Then plot the corresponding vertex of the image: 3 right × 2 = 6 right; 2 up × 2 = 4 up.

4 Use the scale factor to find the length of each side on the new shape: base of shape A is 1, so the new base will be 1 × 2 = 2.

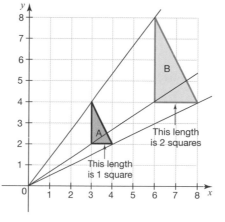

5 Draw the rest of the shape.

Finding the centre of enlargement and the scale factor

To find the centre and scale factor for an enlargement:

1 Draw a straight line from a vertex on the original shape through the corresponding vertex on the image.

2 Repeat this for two other vertices.

③ The centre of enlargement is where the lines intersect.

④ Use a convenient length on the original shape and measure the corresponding length on the enlarged shape:

$$\text{scale factor} = \frac{\text{length of side on image}}{\text{length of side on object}}$$

If the scale factor is greater than 1 then the shape after the enlargement is bigger and if the scale factor is a fraction, then the shape after the enlargement is smaller.

Enlargements with negative scale factors

If the enlargement is the other side of the centre of enlargement, the scale factor is negative.

The enlargement of A to B has scale factor -1 with centre of enlargement (0, 0).

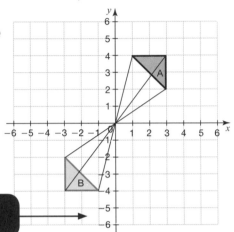

It is called an enlargement even if the shape stays the same size.

DO IT!

Summarise on a revision card the information you need to give to describe each type of transformation.

✓ CHECKIT!

1 Three coordinates, $A(2, 4)$, $B(6, 4)$ and $C(4, 6)$, are joined to make triangle ABC.

Triangle DEF is an enlargement of triangle ABC with scale factor $\frac{1}{2}$ and centre of enlargement (0, 0).

a Draw triangle ABC on a grid, with x and y from 0 to 8.

b Draw DEF on the same grid.

c Describe the single transformation that maps triangle DEF back on to ABC.

2 Triangle P has vertices (1, 3), (4, 3) and (4, −1).

a Draw triangle P on a grid, with x and y from −8 to 8.

b Reflect triangle P in the line $x = -1.5$.

3 Triangle A has vertices (5, 2), (4, 4) and (7, 2).

a Draw triangle A on a grid, with x and y from −8 to 8.

b Enlarge shape A by scale factor −2 with centre of enlargement (2, 1). Label the image of the enlargement B.

4 Describe the single transformation that maps triangle P onto triangle Q.

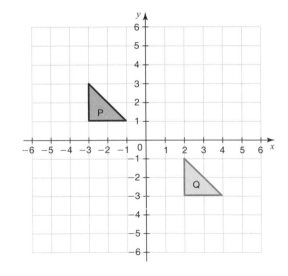

Invariance and combined transformations

Invariance

When combinations of translations, rotations and reflections are applied to a shape, sometimes the positions of all or some of the points on the shape stay the same. These points are said to be **invariant**.

> All the other vertices change but A is invariant as it is the centre of rotation.

WORKIT!

OABC is a square.

a *OABC* is reflected in the line $x = 2$. How many invariant vertices on the square are there?

As vertices A and B lie on the mirror line they will be invariant.

Hence there are two invariant vertices.

b *OABC* is rotated 90° clockwise about the point *A*. How many invariant vertices are there?

There is one invariant vertex, A. ◄

c *OABC* is reflected in the line $y = x$. Give the coordinates of the vertices that are invariant after this reflection.

The invariant vertices are O (0, 0) and B (2, 2). ◄

> Vertices O and B lie on the line $y = x$, which is the mirror line, so both these vertices are invariant.

Combinations of translations, rotations and reflections

When you are told to apply a combination of translations, rotations or reflections to a shape, first transform the original shape and then transform the image to give the final image.

WORKIT!

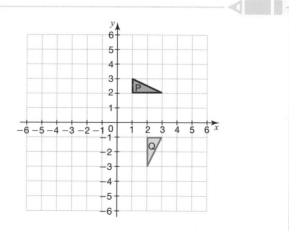

a Rotate triangle P through 90° clockwise about the origin. Label the image Q.

See diagram on right.

b Reflect image Q in the *y*-axis. Label the image R.

See diagram on next page.

c Describe the single transformation that maps triangle P to triangle R.

Reflection in the line $y = -x$

You can see that R is a reflection of P.

Notice all the lines from vertices on P to the line $y = -x$ are the same length as the lines from corresponding vertices on R.

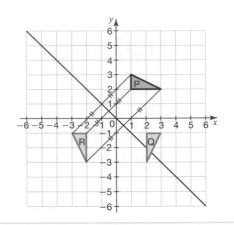

DOIT!

Draw a simple shape and perform two translations on it. Describe the single transformation that has the same effect.

CHECKIT!

1 Shape X has vertices (1, 1), (1, 2), (3, 2) and (4, 1).

a Draw shape X on a grid, with x and y from −5 to 5.

b Reflect shape X in the line $y = -1$. Label the image Y.

c Rotate shape Y through 180° about (0, −1). Label the image Z.

d Describe the single transformation that maps shape X to shape Z.

2 Triangle ABC has vertices A(1, 4), B(3, 4) and C(0, 6).

a Draw triangle ABC on a grid, with x and y from 0 to 6.

b Reflect triangle ABC in the line $x = 3$.

c Write down the coordinates of the invariant point during the reflection.

3 The graph below shows shape A.

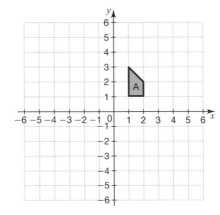

a Rotate shape A 180° about the origin and label it B.

b Shape B is reflected in the y-axis to give shape C. Draw shape C on the graph.

c Describe the single transformation that would be applied to shape C to give shape A.

3D shapes

It is important to know the properties of 3D shapes as you may be asked questions based on these. For example, if you are asked to find the surface area of a square-based pyramid you need to know it has 5 faces: 4 triangles and a square.

Shape		Faces	Edges	Vertices
Cube		6	12	8
Cuboid		6	12	8
Triangular prism		5	9	6
Hexagonal prism		8	18	12
Cylinder		3	2	0
Square-based pyramid		5	8	5

Vertex: a corner (plural is vertices).

Face: a flat exterior surface.

Edge: the line where two faces join.

A **plane** is a flat surface. The diagram shows a plane drawn between vertices *A*, *D*, *H* and *E*.

2 flat faces and 1 curved face.

Shape		Faces	Edges	Vertices
Triangular-based pyramid		4	6	4
Cone		2	1	1
Sphere		1	0	0

A triangular-based pyramid is also called a **tetrahedron**.

The top of a cone is called an apex. This may also be referred to as the vertex.

A plane of symmetry is like a line of symmetry for 3D shapes.

DOIT!

Work out how many planes of symmetry there are in each of the 3D shapes in this table.

CHECKIT!

1 Copy and complete the table.

Shape	Number of vertices	Number of faces	Number of edges
Triangular-based pyramid			
Cone			
Cuboid			
Hexagonal prism			

2 Show that for an octagonal prism, $V + F - E = 2$, where V is the number of vertices, F is the number of faces and E is the number of edges.

Parts of a circle

To answer questions about circles you need to be familiar with all the relevant terms relating to them.

The perimeter of a circle is called the **circumference**. Part of the circumference is called an **arc**.

The **radius** of a circle is the distance from the centre (O) to the edge. It is half of the **diameter**.

A **tangent** is a line that just touches the circumference of a circle.

A **chord** is any straight line across the circle that is not a diameter.

A **segment** is any area cut off by a chord.

A **sector** is a pie slice formed by two radii of the circle.

> The shaded **segment** and **sector** are called the minor segment and minor sector. The unshaded, larger areas are the major segment and major sector.

WORKIT!

Describe what this diagram of a circle shows using the relevant terms.

The perpendicular bisector of a chord of a circle passes through the centre of the circle.

AB is a chord. OA and OB are the same length as they are both radii of the circle.

Triangle OAB is an isosceles triangle so the perpendicular from O is a line of symmetry.

The perpendicular bisects the chord: AM = MB.

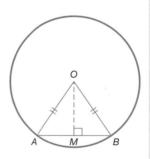

DOIT!

Create matching pairs of cards with the terms and diagrams for parts of a circle. Practise matching the cards.

CHECKIT!

1 Give the names for the labelled parts of a circle (a–e).

Circle theorems

SNAPIT!   Circle theorems

There are lots of circle theorems to remember. Take a picture of each one to help you revise.

Circle theorems are used to work out angles in circles. Solving these types of question takes practice in order to identify the theorem you need to use. The key is to be able to choose the relevant theorems when there are lots of angles in the diagram.

The angle in a semicircle is a right angle.

Make sure that the line you think is the diameter does pass through the centre of the circle.

The angle at the centre is twice the angle at the circumference on the same arc.

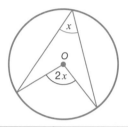

Angles at the circumference on the same arc (or equal arcs) are equal.

In the diagram, the lines making both angles start and finish on the same arc.

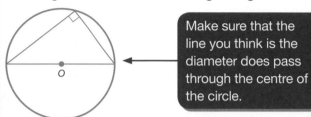

You may need to look carefully to see when to use this. Extra lines drawn on the diagram can make it harder to spot.

Opposite angles of a **cyclic quadrilateral** are **supplementary** (they add up to 180°):
$a + b = 180°$

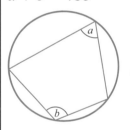

A cyclic quadrilateral has all four vertices on the circumference of a circle.

Alternate segment theorem: if *AB* is the chord of a circle and a tangent *XY* is drawn at *A,* and *C* is any point on the arc *AB*, then angle *ACB* and angle *BAY* are equal.

The angle between the chord and the tangent equals the angle formed at point C in the other (alternate) segment.

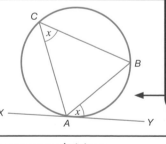

This can be the hardest theorem to spot, so needs practice. In the exam, give the reason as 'alternate segment theorem'.

The tangent at a point on a circle makes a right angle with the radius drawn through the point.

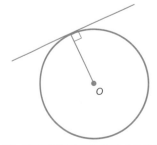

Tangents from the same point to a circle are the same length.

NAILIT!

As well as using circle theorems, you must apply other geometrical properties. So look out for isosceles triangles, parallel lines, etc.

WORKIT!

A, B, C and D are points on the circumference of a circle with centre O. Angle AOC = x.

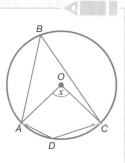

a Find the size of angle ABC in terms of x.

angle ABC = $\frac{x}{2}$ (angle at the centre is twice the angle at the circumference on the same arc)

b Find the size of angle ADC in terms of x.

angle ABC + angle ADC = 180°
(ABCD is a cyclic quadrilateral)

angle ADC = 180 − angle ABC = $\left(180 - \frac{x}{2}\right)°$

NAILIT!

Always give the reason for your answer.

WORKIT!

AB is the diameter of the circle. A tangent XY to the circle is drawn at point B.

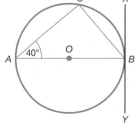

a Write down the size of angle ACB, giving a reason for your answer.

angle ACB = 90°   (angle in a semicircle is a right angle)

b Write down the size of angle CBX, giving a reason for your answer.

angle CBX = angle BAC   (alternate segment theorem)
= 40°

c Write down the size of angle ABY, giving a reason for your answer.

angle ABY = 90°   (angle between tangent and diameter)

SNAPIT!

Take photos of any questions you see with unusual diagrams, showing your solution, too.

DOIT!

Write down a list of the theorems and practise drawing a diagram for each one.

✓ CHECKIT!

1 ABCD is a cyclic quadrilateral and O is the centre of the circle.

Angle YAB = 30°

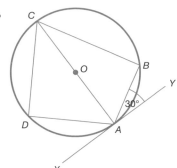

a Work out the size of angle ACB, giving a reason for your answer.

b Write down the size of angle ABC, giving a reason for your answer.

c Write down the size of angle ADC, giving a reason for your answer.

2 *A*, *B* and *C* are points on the circumference of a circle. A tangent *AX* is drawn at *A*.

Angle *AXO* = 30°.

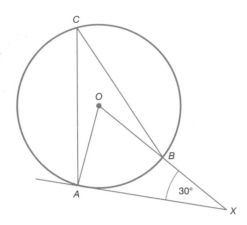

a State the size of angle *OAX*.

b State, with a reason, the size of angle *AOX*.

c State, with a reason, the size of angle *ACB*.

You must give a reason for each stage of your working.

3 *A*, *B*, *C* and *D* are points on a circle. A tangent is drawn to the circle at *D*.

Angle *ACB* = 40° and angle *EDA* = 50°.

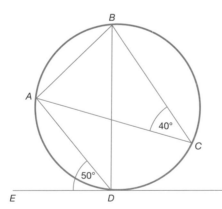

a Write down, with a reason, the size of angle *ADB*.

b Calculate angle *BAD*.

You must give a reason for each stage of your working.

4 *A*, *B*, *C* and *D* are points on the circumference of a circle with centre *O*.

Angle *ADB* = 46°

Angle *DBC* = 50°

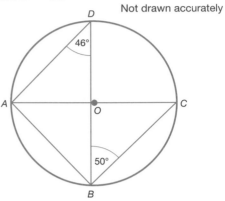

Work out the size of angle *BAC* giving reasons for your answer.

NAILIT!

Diagrams may be marked 'not drawn accurately' or 'not to scale'. This means the sizes and proportions shown are not exact – angles or lengths that look similar may not be. Use the information you are given to work them out.

5 *A*, *B*, *C* and *D* are points on the circumference of a circle. *ABC* is an equilateral triangle and *PQ* is a tangent to the circle at *B*.

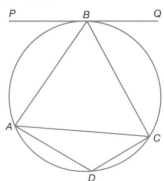

a Work out the size of angle *ADC*.

b Work out the size of angle *CBQ*.

You must give a reason for each stage of your working.

Projections

Projections are views of a solid shape from different directions such as:

- **front elevation** – what you would see directly in front of the solid
- **side elevations** – view from either the left or the right ◄
- **plan** – view from directly above.

> You will be told in the question whether both or just one side is required.

> In the exam, you will be supplied with the grid.

WORKIT!

The solid shape shown is made up from 11 identical cubes.

Draw on a grid the side elevation from L, the side elevation from R and the plan view.

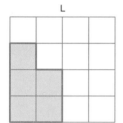

> **NAILIT!**
>
> Work from left to right of the view and shade in the squares as you go.

> **DOIT!**
>
> Build some shapes from cubes and draw two elevations and a plan view of each one.

✓ CHECKIT!

1 This solid shape is made up of 8 identical cubes.

On a grid, draw:

a the side elevation from the left

b the front elevation

c the plan view.

2 The plan, front elevation and side elevation of a prism are drawn on a centimetre grid.

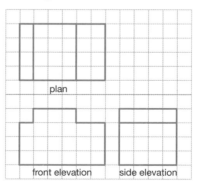

a Draw a sketch of the prism.

b Write the dimensions of the prism on your sketch.

Bearings

NAILIT!

You must be able to draw lines in the compass directions north, east, south and west.

NAILIT!

The 'from' part in the question is important because it tells you where you are starting from. 'From point A' means you should measure the bearing from this point.

NAILIT!

Add north lines through all the points and mark any angles mentioned in the question that are not included on the diagram.

DOIT!

Write the main eight compass directions (N, NE, E, ...) as bearings.

Bearings are always:

- measured from due north
- measured in a clockwise direction
- written as three figures (for example, a bearing of 5° would be 005° and a bearing of 32.5° would be 033°).

East is a bearing of 090°, south 180° and west 270°.

NAILIT!

You often need to look for alternate angles when you draw diagrams showing bearings between two or more points.

WORKIT!

> Draw a diagram.

Point B is on a bearing of 100° from point A.

What is the bearing of point A from point B?

From the diagram, θ = 180 + 100 = 280°

The bearing of point A from point B is 280°

> The bearing from B to A is the clockwise angle θ.

> North lines are always parallel so you can use these to identify alternate angles.

NAILIT!

If the question does not include a diagram, draw one to help you understand the problem.

CHECKIT!

1 Ship S sets off on a straight course of 245° for port P.

What is the bearing of P from S?

2 The diagram shows the positions of three wind turbines, P, Q and R.

The bearing of Q from P is 045°.

The bearing of R from Q is 140°.

a Calculate the bearing of P from Q.

b Calculate the bearing of Q from R.

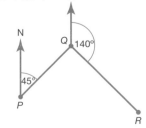

Pythagoras' theorem

Pythagoras' theorem states that the square of the hypotenuse of any right-angled triangle is equal to the lengths of the other two sides squared and added together.

SNAPIT! Pythagoras' theorem

$$c^2 = a^2 + b^2$$

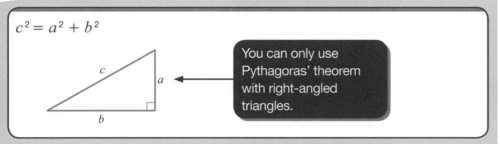

You can only use Pythagoras' theorem with right-angled triangles.

> **Memorise:** the square on the hypotenuse is equal to the sum of the squares on the other two sides.

NAILIT!

The **area** of a triangle is covered on page 153.

NAILIT!

You need to learn the formula for Pythagoras' theorem, as you will not be given it in the exam.

WORKIT!

An isosceles triangle has base 8 cm and the pair of equal sides of length 10 cm. Find, to 3 significant figures

a the perpendicular height of the triangle

Using Pythagoras' theorem $10^2 = h^2 + 4^2$

$$100 = h^2 + 16$$

$$84 = h^2$$

$$h = \sqrt{84} = 9.17\,\text{cm} \text{ (to 3 s.f.)}$$

10 cm h 4 cm

b the area of the triangle.

Area $= \frac{1}{2} \times$ base $\times$ height

$= \frac{1}{2} \times 8 \times 9.17$

$= 36.7\,\text{cm}^2$ (to 3 s.f.)

Draw a diagram. The perpendicular height of an isosceles triangle cuts the base into two equal lengths.

NAILIT!

If you are asked for an 'exact answer', leave your answer in surd form.

DOIT!

Work out a formula for finding the distance between two points whose coordinates are given.

NAILIT!

Pythagoras' theorem can occur in many questions on different topics, for example finding the distance between two points whose coordinates are given.

SNAPIT!

Take photos of right-angled triangles you see around you, and use them to create your own questions.

CHECKIT!

1 Find length x in these triangles, giving your answers to 1 decimal place.

a

5 cm
9 cm
x

b

10.2 cm
x
13.5 cm

2 The lengths of the sides of an isosceles triangle are 12 cm, 12 cm and 10 cm. Find the perpendicular height of the triangle and hence its area, giving your answer to 2 decimal places.

3 Triangles *ABC* and *ACD* are right-angled triangles.

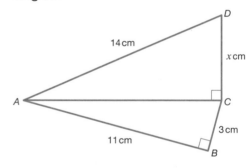

D
14 cm
x cm
A
C
3 cm
11 cm
B

Work out the value of x, correct to 2 decimal places.

4 The isosceles triangle shown below has base $AB = 14$ cm.

The area of the triangle *ABC* is 90 cm².

Calculate the length of *AC* giving your answer to 3 significant figures.

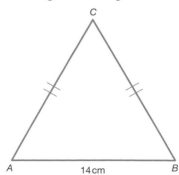

C
A
14 cm
B

Area of 2D shapes

SNAP IT!   Area and volume

There are lots of formulae to remember for area and volume.
Take a picture of each one to help you revise.

NAIL IT!

Sometimes you need to use Pythagoras' theorem to find the height of a trapezium.

Area of triangle $= \frac{1}{2} \times$ base $\times$ perpendicular height	Area of parallelogram $=$ base $\times$ perpendicular height	Area of trapezium $= \frac{1}{2}(a + b)h$	Circumference of circle $= 2\pi r = \pi D$ Area of circle $= \pi r^2$

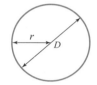

NAIL IT!

The Greek letter π equals the area of a circle with radius 1 unit and has the value 3.14159... It is an irrational number (a non-repeating, non-terminating decimal). Check that you know which key to use on your calculator.

DO IT!

Write each formula and its shape on pairs of cards. Practise matching the cards.

CHECK IT!

1 A wooden floor is going to be put in the office shown by this plan. The floor costs £38 per m². How much will the new floor cost?

9 m
4 m
5 m
12 m

2 The diagram shows a square with a perimeter of 20 cm.

2.5 cm
2.5 cm

Without using a calculator, work out the shaded area as a proportion of the area of the square. Give your answer as a fraction.

3 Shape $ABCDE$ consists of a semicircle with diameter BD and a rectangle $ABDE$.

$BD = 2x$ and $AB = 4x$.

a Prove that the area of the shape is $x^2\left(8 + \frac{\pi}{2}\right)$.

b Find the exact perimeter of the shape $ABCDE$ in terms of x.

Volume and surface area of 3D shapes

SNAPIT! **Volume**

There are lots of formulae for area and volume. You will be given some of these formulae in the question, but take a picture of each one to help you revise.

Volume

Volume of cuboid = length × width × height

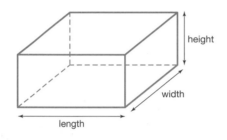

Volume of cylinder $= \pi r^2 h$

A cylinder is a prism with a circular cross-section, so the area can be found by multiplying the cross-sectional area by the height/length.

Volume of prism = area of cross-section × length

A prism has the same cross-sectional shape along its length.

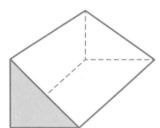

Volume of a cone $= \frac{1}{3}\pi r^2 h$

Volume of sphere $= \frac{4}{3}\pi r^3$

WORKIT!

1 *ABCDEF* is a prism with an equilateral triangle cross-section of side 3 cm. The length of the prism is 12 cm.

> Volume of prism = area of cross-section × length. First find the area of the cross-section.

Work out the volume of the prism. Give your answer to 2 significant figures.

Using Pythagoras' theorem, $3^2 = h^2 + 1.5^2$

$9 = h^2 + 2.25$

$h = 2.60$ cm (to 2 d.p.)

> To find the area of the triangle you need to find the perpendicular height, h, first.

Area of triangle $= \frac{1}{2} \times$ base $\times$ height

$= \frac{1}{2} \times 3 \times 2.60 = 3.9\,cm^2$ (to 2 s.f.)

Volume of prism = area of cross-section × length

$= 3.9 \times 12 = 47\,cm^3$ (to 2 s.f.)

2 100 steel ball bearings of radius 1 cm are melted down and made into a cuboid with a base area of 60 cm². Calculate the height of the cuboid in cm, giving your answer correct to 3 significant figures.

Volume of one ball bearing $= \frac{4}{3}\pi r^3 = \frac{4}{3}\pi \times 1^3 = 4.1888\,cm^3$

> First find the volume of one sphere.

Volume of 100 ball bearings $= 100 \times 4.1888 = 418.88\,cm^3$

Volume of cuboid = area of base × height

> A cuboid is an example of a prism with a rectangular cross-section.

$418.88 = 60 \times h$

$h = \frac{418.88}{60}\,cm = 6.98\,cm$ (to 3 s.f.)

Surface area

Curved surface area of cylinder $= 2\pi rh$ Total surface area of cylinder $= 2\pi rh + 2\pi r^2$	Curved surface area of cone $= \pi rl$ Total surface area of cone $= \pi rl + \pi r^2$	Surface area of sphere $= 4\pi r^2$

> Note that the formula for the curved surface area uses the slant height, but the volume uses the perpendicular height.

NAILIT!

Sometimes you may be given the perpendicular height and the radius. You need to use Pythagoras' theorem to find the slant height: $r^2 + h^2 = l^2$.

Volume and surface area of frustums

A frustum is the solid that remains when the top is cut off a cone, parallel to its circular base.

Volume of frustum = volume of original cone −
volume of removed top cone

$$= \tfrac{1}{3}\pi R^2 H - \tfrac{1}{3}\pi r^2 h$$

Surface area of frustum = curved surface area of original cone

− curved surface area of removed top cone

+ area of top circle

+ area of bottom circle

$$= \pi RL - \pi rl + \pi r^2 + \pi R^2$$

NAILIT!

If you are asked to find the area of the curved surface it does not include the circular top and bottom.

WORKIT!

A cone has height 8 cm and base radius 6 cm.

a Work out the surface area of the cone. Give your answer in terms of π.

First work out the slant height, l, of the cone.

Using Pythagoras' theorem:

$l^2 = r^2 + h^2$

$= 6^2 + 8^2$

$= 36 + 64$

$= 100$

$l = 10$ cm

Remember to include the area of the base of the cone.

Surface area = curved surface area + base area

$= \pi rl + \pi r^2$

$= \pi \times 6 \times 10 + \pi \times 6^2$

Leave your answer in terms of π.

$= 96\pi$ cm^2

b A cone of radius 1.5 cm and height 2 cm is removed from the top of the cone to create a frustum.

Work out the volume of the frustum. Give your answer to 3 significant figures.

Volume = volume of large cone − volume of small cone

$= \tfrac{1}{3} \times \pi \times 6^2 \times 8 - \tfrac{1}{3} \times \pi \times 1.5^2 \times 2$

$= 297$ cm^3 (to 3 s.f.) ◄── Check the accuracy asked for in the answer.

WORKIT!

A solid sphere of radius 30 cm is cut in half to form two identical solids. Find the surface area of one of the solids, giving your answer in terms of π.

Curved surface area of hemisphere $= \dfrac{4\pi r^2}{2} = 2\pi r^2 = 2 \times \pi \times 30^2$

$= 1800\pi \text{ cm}^2$

Area of circular base $\pi r^2 = \pi \times 30^2 = 900\pi \text{ cm}^2$

Total surface area of solid $= 1800\pi + 900\pi = 2700\pi \text{ cm}^2$

> The surface area is the curved surface area of the hemisphere plus the area of the circular base.

> Remember to leave answers in terms of π.

DOIT!

Write each formula and its shape on pairs of cards. Practise matching the cards.

CHECKIT!

1 A cylinder of radius a m and height $6a$ m is reformed into a thin cylindrical rod with radius r and length 4 m. Write an expression for the radius of the rod, r, in terms of a.

2 The diagram shows the cross-section of a rectangular swimming pool.

a Given that its volume is a prism, find the cross-sectional area shaded in the diagram.

b Calculate the volume of water in the swimming pool in m³.

The swimming pool is filled using two hoses. Each hose supplies a volume of water at 0.05 m³ per minute.

c Calculate the number of hours it would take to fill the pool with water. Give your answer to the nearest hour.

3 A circus tent is in the shape of a cone on a cylinder. The radius of the base is 15 m, the height of the cylinder is 5 m and the perpendicular height of the cone is 4 m.

Find the surface area of fabric used for the tent, correct to 3 significant figures. Assume that the tent has no floor.

4 A cup is made by removing a cone of radius 3.5 cm and height 14 cm from a larger cone of radius 5 cm and height 20 cm.

Work out the volume of liquid the cup can hold.

Give your answer in terms of π.

Trigonometric ratios

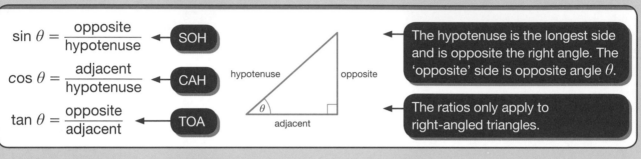

$$\sin \theta = \frac{\text{opposite}}{\text{hypotenuse}} \quad \leftarrow \boxed{\text{SOH}}$$

$$\cos \theta = \frac{\text{adjacent}}{\text{hypotenuse}} \quad \leftarrow \boxed{\text{CAH}}$$

$$\tan \theta = \frac{\text{opposite}}{\text{adjacent}} \quad \leftarrow \boxed{\text{TOA}}$$

The hypotenuse is the longest side and is opposite the right angle. The 'opposite' side is opposite angle θ.

The ratios only apply to right-angled triangles.

NAILIT!

These ratios will not be given in the exam so you need to remember them.

Use the mnemonic SOH CAH TOA to help.

Finding a length

Here's how to use the trigonometric ratios to find an unknown length in a right-angled triangle when one side and one angle are known.

1. Identify which side is known.

2. Work out which trigonometric ratio to use from SOH CAH TOA.

3. Write the trigonometric ratio and substitute the values.

4. Solve the equation for the unknown length.

WORKIT!

Find the length of the side marked x, giving your answer to 3 significant figures.

$$\tan \theta = \frac{opposite}{adjacent}$$

The sides are O and A so use TOA.

$$\tan 30° = \frac{6}{x}$$

$$x \tan 30° = 6$$

$$x = \frac{6}{\tan 30°} = 10.4 \, cm \, \text{(to 3 s.f.)}$$

Write the values in $\tan \theta = \frac{\text{opposite}}{\text{adjacent}}$ x is in the denominator of the fraction.

Finding an angle

Similarly, you can use the trigonometric ratios to find an angle of a right-angled triangle when the lengths of two sides are known. First find the sin, cos or tan. Then find the angle.

WORKIT!

1 Find the size of the angle θ, giving your answer to 1 decimal place.

$\cos \theta = \dfrac{\text{adjacent}}{\text{hypotenuse}}$ ← The sides are A and H so use CAH.

$= \dfrac{8}{10} = 0.8$

$\theta = \cos^{-1}(0.8)$

$= 36.9°$ (to 1 d.p.)

10 cm

8 cm

θ

2 Triangle *ABC* has an area of 25 cm². Side *AB* is 10 cm.

a Find the perpendicular height, *h*.

$\text{Area} = \dfrac{1}{2} \times \text{base} \times \text{perpendicular height}$

$25 = \dfrac{1}{2} \times 10 \times h$

$h = 5 \text{cm}$

C

A 30° 10 cm *B*

h

b Calculate the length *AC*.

$\sin 30° = \dfrac{5}{AC}$ ← Substitute values into $\sin \theta = \dfrac{\text{opposite}}{\text{hypotenuse}}$.

$AC = \dfrac{5}{\sin 30°} = \dfrac{5}{0.5} = 10 \text{cm}$

NAILIT!

Check that you know how to find inverse sine, cosine and tangent on your calculator. The key may be shown as

sin⁻¹   or   arcsin

Solving problems in 3D

To solve 3D problems, you need to identify right-angled triangles and then use these with Pythagoras' theorem and the trigonometric ratios.

WORKIT!

The diagram shows the internal dimensions of a rectangular box.

a Find the length of the longest straight rod that will just fit inside the box.

The longest diagonal will be one of the lengths like *AD*. →

$AC^2 = 60^2 + 30^2 = 4500$

Use Pythagoras' theorem with triangle *ABC* to find AC^2. →

$AD^2 = 4500 + 20^2 = 4900$ ←

$AD = \sqrt{4900} = 70 \text{cm}$

Longest length of rod is 70 cm.

Angle *ACD* is 90° so use Pythagoras' theorem with triangle *ACD* to find *AD*.

b Work out the angle that the rod makes with the base of the box.

Angle *CAD* is the angle with the base of the box. →

Let angle $CAD = \theta$

$\sin \theta = \dfrac{\text{opposite}}{\text{hypotenuse}} = \dfrac{20}{70} = 0.2857$ ←

Use sides *CD* and *AD* with sine.

$\theta = \sin^{-1}(0.2857) = 16.6°$ (to 1 d.p.)

The rod makes an angle of 16.6° with the base of the box.

D

20 cm

C

30 cm

A   60 cm   *B*

NAILIT!

In 3D questions, you may find it helpful to draw just the triangle you are working on.

DOIT!

Look up different mnemonics for remembering the ratios. Choose one you find easy to remember.

CHECKIT!

1 Find the lengths of the sides marked x. Give your answers to 2 decimal places.

a

b

2 Find angle θ for each of these triangles. Give your answers to the nearest 0.1°.

a

b

3 In terms of a, b and c, prove that for $0° < \theta < 90°$, $\dfrac{\sin\theta}{\cos\theta} = \tan\theta$.

4 The diagram shows a cuboid box with edges $AB = 10\,cm$, $AD = 7\,cm$ and $BC = 6\,cm$.

a Calculate the length of CD, correct to 2 decimal places.

b Work out angle BCD, correct to 1 decimal place.

c Point P is the midpoint of BD. Calculate the length of PC, correct to 2 decimal places.

Exact values of sin, cos and tan

You already know the values of the sine and cosine of 0° and 90° and of tan 0° from the graphs.

The values for 30°, 45° and 60° can be determined by drawing triangles, working out the lengths of the sides that aren't known and then using trigonometry.

> You can find the graphs of sine, cosine and tangent on page 85.

Sine, cosine and tangent of 45°

The exact values can be worked out by drawing a right-angled triangle with angles of 45° and short sides of 1 unit.

$$\sin 45° = \frac{\text{opposite}}{\text{hypotenuse}} = \frac{1}{\sqrt{2}}$$

> Substitute the lengths of the sides in the trigonometric ratios.

$$\cos 45° = \frac{\text{adjacent}}{\text{hypotenuse}} = \frac{1}{\sqrt{2}}$$

$$\tan 45° = \frac{\text{opposite}}{\text{adjacent}} = \frac{1}{1} = 1$$

> Using Pythagoras' theorem, hypotenuse $= \sqrt{1^2 + 1^2} = \sqrt{2}$

These are all exact values. For example, entering cos 45° into a calculator gives 0.7071067811…, which cannot be expressed as a fraction. In calculations where answers are given to a number of decimal places or significant figures, the answers are not exact values. Numbers expressed in terms of surds are exact.

> **NAILIT!**
>
> If you are asked to find the exact value use surds instead of decimals.

Sine, cosine and tangent of 30° and 60°

These exact values can be worked out by using an equilateral triangle of side length 2. Bisecting one of the sides then gives a right-angled triangle (A).

$$\sin 30° = \frac{1}{2} \qquad \sin 60° = \frac{\sqrt{3}}{2}$$

$$\cos 30° = \frac{\sqrt{3}}{2} \qquad \cos 60° = \frac{1}{2}$$

$$\tan 30° = \frac{1}{\sqrt{3}} \qquad \tan 60° = \sqrt{3}$$

> All the angles in the equilateral triangle are 60°.

> Bisecting the triangle gives a right-angled triangle with angles 60° and 30°.

> Using Pythagoras' theorem, base $= \sqrt{2^2 - 1^2} = \sqrt{3}$

> **DOIT!**
>
> Draw a table of the values of sin, cos and tan of 0°, 30°, 45°, 60° and 90°.

CHECKIT!

1 Show that $\sqrt{3} \tan 30° + \cos 60°$ can be written as $\frac{a}{b}$, where a and b are integers.

2 Show that $\sin^2 30° + \cos^2 30° = 1$.

Sectors of circles

SNAPIT! Sectors

For a sector of radius r and angle θ:

- length of arc: $l = \dfrac{\theta}{360} \times 2\pi r$

- area of sector: $A = \dfrac{\theta}{360} \times \pi r^2$

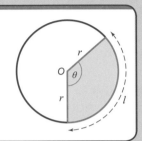

You can work out the arc length of a sector of angle θ degrees by multiplying the circumference of the circle by $\dfrac{\theta}{360}$. Similarly, you can work out the area of a sector of angle θ degrees by multiplying the area of the circle by $\dfrac{\theta}{360}$.

WORKIT!

Judy is planning a wedding to be attended by 57 guests. The tables have a diameter of 2.3 m. Each guest needs a distance of 70 cm along an arc around the table.

Calculate the maximum number of guests that can be seated at each table.

Circumference of table $= \pi D = \pi \times 2.3 = 7.23\text{ m}$

$7.23\text{ m} = 723\text{ cm}$

> Make sure that both numbers are in the same units.

Number of guests at each table $= \dfrac{723}{70} = 10.3$

> The number of guests must be a whole number.

Maximum number at a table $= 10$

WORKIT!

The diagram shows a sector with radius 6 cm. Its area is 24 cm².

a Work out angle x, giving your answer to 1 decimal place.

> Write down the formula for the area of a sector.

$A = \dfrac{x}{360}\pi r^2$

$24 = \dfrac{x}{360} \times \pi \times 6^2$

> Substitute the values you know.

> Cancel this before calculating.

$x = \dfrac{24 \times 360}{\pi \times 36} = 76.4°$ (to 1 d.p.)

b Calculate the length of the arc, l, giving your answer to the nearest whole number.

> Write down the formula for the area of a sector.

$l = \dfrac{x}{360} \times 2\pi r = \dfrac{76.4}{360} \times 2\pi \times 6 = 8\text{ cm}$ (to nearest whole number)

DO IT!

Make a revision poster for the length of an arc and the perimeter and area of a sector.

CHECKIT!

1 *AB* is the arc of a circle of radius 12 cm. The length of the arc *AB* is 10 cm.

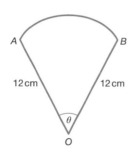

Calculate the size of angle θ to 1 decimal place.

2 This logo is made of two identical sectors and an isosceles triangle. The sectors have radius 2 cm and angle 40°. The base of the isosceles triangle is 1.2 cm.

Work out

a the perimeter of the logo

b the area of the logo.

Give your answers to 3 significant figures.

3 The area of sector *OAB* is 25π cm².

Find the radius *OA* to 1 decimal place.

4 A circular piece of paper has a sector removed to leave the shaded shape shown below.

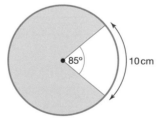

Calculate the area of the shaded shape. Give your answer to the nearest whole number.

Sine and cosine rules

The **sine rule** and **cosine rule** can be used with any triangle, not just those containing a right angle.

The angles are denoted by the letters A, B and C and the lengths of the sides opposite these angles by a, b and c.

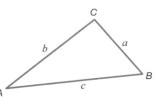

SNAP IT! Sine and cosine rules

The sine rule: $\dfrac{a}{\sin A} = \dfrac{b}{\sin B} = \dfrac{c}{\sin C}$

The cosine rule: $a^2 = b^2 + c^2 - 2bc\cos A$

DO IT!

Using the diagram above and what you already know about triangles, see whether you can prove that these rules are true.

The area of a triangle

If you don't know the height of a triangle but you do know the lengths of two sides and the included angle, you can find its area using the formula below.

SNAP IT! Area of any triangle

Area of triangle $= \dfrac{1}{2}ab\sin C$ ◄

This formula works for all triangles, but for areas of right-angled triangles, use the formula $A = \dfrac{1}{2} \times$ base $\times$ height.

Warning: be careful when using this formula to find the angle from the area and two sides of a triangle. For example, $\sin C = \dfrac{1}{2}$ can have two solutions: 30° and 150° so the triangle could be obtuse-angled. If the triangle has been drawn or other angles in the triangle are known, then the obtuse angle might not be a possible solution. Check the question carefully.

WORKIT!

1 In triangle PQR, $PR = 8$ cm, $QR = 12$ cm and angle $PRQ = 150°$.

a Find the area of triangle PQR.

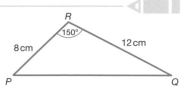

Area $= \dfrac{1}{2}ab\sin C = \dfrac{1}{2} \times 12 \times 8 \times \sin 150° = 24$ cm^2

Always write out the formula and then substitute the numbers you know.

b Find the length of *PQ*, correct to 1 decimal place.

Using the cosine rule:

$a^2 = b^2 + c^2 - 2bc\cos A$

$= 12^2 + 8^2 - 2 \times 12 \times 8 \times \dfrac{-\sqrt{3}}{2}$ ← cos 150° = −cos 30°

$= 374.2769$

$a = 19.3462$

$PQ = 19.3\,cm$ (to 1 d.p.)

NAILIT!

Don't round off answers to the required number of decimal places until the final answer.

2 In triangle *ABC*, *AB* = 8 cm, *BC* = 15 cm and angle *ABC* = 60°.

a Calculate the length of side *AC*.

Using the cosine rule:

$a^2 = b^2 + c^2 - 2bc\cos A$

$AC^2 = 8^2 + 15^2 - 2 \times 8 \times 15 \times \cos 60°$

$= 64 + 225 - 240 \times \dfrac{1}{2}$

$= 169$

$AC = 13\,cm$

DOIT!

Draw a right-angled triangle with sides *a*, *b* and *c*. What happens when you apply the sine and cosine rules and the formula for the area of a triangle to this triangle?

b Find the size of angle θ, giving your answer to 1 decimal place.

Using the sine rule:

$\dfrac{b}{\sin B} = \dfrac{c}{\sin C}$

$\dfrac{13}{\sin 60°} = \dfrac{8}{\sin \theta}$

$\sin \theta = \dfrac{8 \times \sin 60°}{13}$

$\theta = 32.2°$ (to 1 d.p.)

✓ CHECK**IT!**

1 In triangle *ABC*, *AB* = 6 cm, *AC* = 10 cm and *BAC* = 150°.

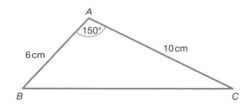

a Find the area of triangle *ABC*.

b Find the length of *BC*, correct to 3 significant figures.

2 A triangle has sides 5 cm, 7 cm and 10 cm. Calculate the smallest angle of the triangle, to the nearest 0.1°.

3 In triangle *XYZ*, angle *XZY* is obtuse.

XY = 9.5 cm and *YZ* = 5.4 cm.

The area of the triangle is 16 cm².

a Find angle *XYZ*, correct to 1 decimal place.

b If you did not know that angle *XZY* was obtuse, how would this affect your answer to part a?

Vectors

Vectors are quantities that have both a **size** and a **direction**. There are a few ways to write them:

- small letters or combinations of small letters written in bold (e.g. 2**a**, **a** + **b**, 3(**a** − **b**))

- two letters with an arrow over them (e.g. $\overrightarrow{OP}$) showing the starting and finishing points of the line, together with the direction (O to P in this case)

- a column vector as shown in the box below.

> $\begin{pmatrix} 3 \\ -1 \end{pmatrix}$ means 3 units horizontally to the right and 1 unit down
>
> $\begin{pmatrix} -4 \\ 5 \end{pmatrix}$ means 4 units horizontally to the left and 5 units up

If $\overrightarrow{OA}$ = **a** and $\overrightarrow{OB}$ = **b**, then the vector from A to B is found by going from A to O and then from O to B. The vector A to O is minus the vector from O to A.

This can be written as $\overrightarrow{AB} = \overrightarrow{AO} + \overrightarrow{OB}$ or −**a** + **b** = **b** − **a**.

If one vector is **a** and the other is −**a**, then the two vectors are the same length but opposite in direction.

If two vectors are parallel, they have the same **vector part**, for example **a** and 2**a**. The numbers in front are called **scalars** and show how long the vector is. 2**a** is therefore twice as long as **a**, and $\frac{2}{3}$**b** is $\frac{2}{3}$ the length of **b**.

2**a** + 2**b** can be written as 2(**a** + **b**). This vector is parallel to and twice the length of **a** + **b**.

The table gives further examples.

> You can think of a vector part as being like a factor. For example, **a** is a factor of 2**a**.

1st vector	2nd vector	Direction?	Length?	Parallel?
4**a**	4**a**	Same	Same	Yes
2**a** + **b**	4**a** + 2**b**	Same	2nd vector is twice as long	Yes
4**a** − **b**	4**a** + **b**	Not the same	Not the same	No
3(**b** − **a**)	2(**b** − **a**)	Same	1st vector is longer with lengths in ratio of 3:2	Yes
−4**a** + 2**b**	**b** − 2**a**	Same	1st vector is twice as long	Yes
−(3**a** − **b**)	3**a** − **b**	Opposite	Same	Yes
$\frac{1}{3}$(**a** − **b**)	**b** − **a**	Opposite	2nd vector is three times as long	Yes

This factorises to 2(2**a** + **b**).

These vectors don't share any factors.

They share the vector part (**b** − **a**).

This vector is the same as 2(**b** − 2**a**).

(**b** − **a**) = −(**a** − **b**)

Calculating with column vectors

You can calculate with column vectors:

- To add $\binom{3}{2}$ and $\binom{4}{-1}$, add the separate elements:

$$\binom{3}{2} + \binom{4}{-1} = \binom{3+4}{2+(-1)} = \binom{7}{1}$$

- To subtract $\binom{4}{-1}$ from $\binom{3}{2}$, subtract the separate elements:

$$\binom{3}{2} - \binom{4}{-1} = \binom{3-4}{2-(-1)} = \binom{-1}{3}$$

- To multiply a vector by a number, multiply each element by the number:

$$5\binom{-2}{4} = \binom{5\times(-2)}{5\times 4} = \binom{-10}{20}$$

WORKIT!

$a = \binom{-5}{3}$ and $b = \binom{7}{1}$

Work out as a column vector

a $a + 2b$

$$a + 2b = \binom{-5}{3} + 2\binom{7}{1} = \binom{-5 + 2\times 7}{3 + 2\times 1} = \binom{9}{5}$$

b $2a - 3b$

$$2a - 3b = 2\binom{-5}{3} - 3\binom{7}{1} = \binom{2\times(-5) - 3\times 7}{2\times 3 - 3\times 1} = \binom{-31}{3}$$

Working with vectors

WORKIT!

Make sure that you use vector $\overrightarrow{PO}$ not $\overrightarrow{OP}$.

1 OPQ is a triangle. M is the midpoint of PQ.
$\overrightarrow{OP} = a$ and $\overrightarrow{OQ} = b$.

a Express $\overrightarrow{PQ}$ in terms of a and b.

$$\overrightarrow{PQ} = \overrightarrow{PO} + \overrightarrow{OQ}$$

$$= -a + b \quad \boxed{\overrightarrow{PO} = -\overrightarrow{OP}.}$$

$$= b - a$$

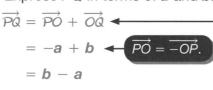

b Express $\overrightarrow{OM}$ in terms of a and b, giving your answer in its simplest form.

$$\overrightarrow{PM} = \tfrac{1}{2}\overrightarrow{PQ} = \tfrac{1}{2}(b - a) \quad \boxed{\text{M is the midpoint of } \overrightarrow{PQ} \text{ so } \overrightarrow{PM} = \tfrac{1}{2}\overrightarrow{PQ}.}$$

$$\overrightarrow{OM} = \overrightarrow{OP} + \overrightarrow{PM}$$

$$= a + \tfrac{1}{2}(b - a)$$

$$= a + \tfrac{1}{2}b - \tfrac{1}{2}a$$

$$= \tfrac{1}{2}(b + a)$$

DOIT!

Write some column vectors on revision cards. Pick two or three cards at a time and make a question from them.

NAILIT!

If it is possible, always take a factor outside the bracket. This helps you to spot any parallel vectors/lines.

NAILIT!

Always take time to redraw the diagram, adding any new points and lines.

M is the midpoint of $\overrightarrow{AB}$, so find vector $\overrightarrow{AB}$ in terms of **a** and **b**.

NAILIT!

If *P* divides vector $\overrightarrow{OA}$ = **a** such that $OP:PA = 1:2$, then $\overrightarrow{OP} = \frac{1}{3}$**a** and $\overrightarrow{PA} = \frac{2}{3}$**a**.

2 *OAB* is a triangle with $\overrightarrow{OA}$ = **a** and $\overrightarrow{OB}$ = **b**.

a M is the midpoint of $\overrightarrow{AB}$ and *Q* is the midpoint of $\overrightarrow{OA}$. Find $\overrightarrow{QM}$.

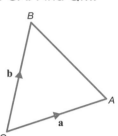

$\overrightarrow{AB} = \overrightarrow{AO} + \overrightarrow{OB} = -a + b = b - a$

$\overrightarrow{AM} = \frac{1}{2}\overrightarrow{AB} = \frac{1}{2}(b - a)$ ◄ *M* is the midpoint of $\overrightarrow{AB}$ so $\overrightarrow{AM} = \frac{1}{2}\overrightarrow{AB}$.

$\overrightarrow{QA} = \frac{1}{2}\overrightarrow{OA} = \frac{1}{2}a$

$\overrightarrow{QM} = \frac{1}{2}\overrightarrow{OA} + \overrightarrow{AM} = \frac{1}{2}a + \frac{1}{2}(b - a)$

$\qquad\qquad = \frac{1}{2}a + \frac{1}{2}b - \frac{1}{2}a$

$\qquad\qquad = \frac{1}{2}b$

b Using your answer to part a, prove that *QM* and *OB* are parallel.

$\overrightarrow{QM}$ and $\overrightarrow{OB}$ have the same vector part, **b**, so they must be parallel.

The same vector part means the vectors are parallel and the scalar shows that $\overrightarrow{QM}$ is half the length of $\overrightarrow{OB}$.

CHECKIT!

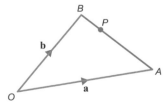

1 $a = \begin{pmatrix} 4 \\ -5 \end{pmatrix}$ and $b = \begin{pmatrix} -2 \\ 3 \end{pmatrix}$

Work out as a column vector

a $b - a$ **b** $3a + 5b$

2 *OAB* is a triangle with $\overrightarrow{OA}$ = **a** and $\overrightarrow{OB}$ = **b**.

a Find $\overrightarrow{AB}$ in terms of **a** and **b**.

P is the point on *AB* such that $AP:PB = 3:2$.

b Find $\overrightarrow{AP}$ in terms of **a** and **b**.

c Find $\overrightarrow{OP}$ in terms of **a** and **b**.

3 *ABC* is a triangle and *ABD* is a straight line.

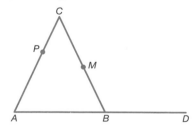

$\overrightarrow{AB}$ = **a**, $\overrightarrow{CP}$ = **b** and $\overrightarrow{PA}$ = 2**b**.

a Find the vector $\overrightarrow{BC}$ in terms of **a** and **b**.

b *M* is the midpoint of *BC* and *B* is the midpoint of *AD*.

Prove that *PMD* is a straight line.

1 A, B, C and D are points on the circumference of a circle. EDF is a tangent to the circle at D.

Angle $ABC = x°$ and angle $ACD = y°$.

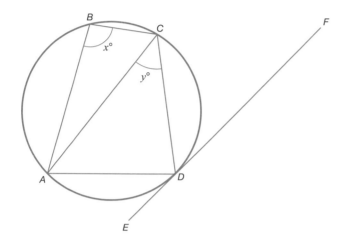

Work out angle CDF in terms of x and y. Give reasons for every stage of your working.

2 Show that $\cos 45° + \sin 60° = \frac{1}{2}(\sqrt{2} + \sqrt{3})$.

3 A right-angled triangle has sides $4x - 3$, $x + 1$ and $3x$.

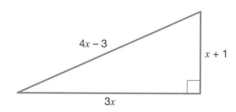

Use Pythagoras' theorem to find the value of x and hence find the lengths of the sides of the triangle.

4 The diagram shows a solid shape made up of seven identical one-centimetre cubes.

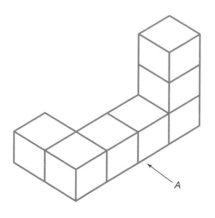

On centimetre grids, draw

a the elevation from A

b the plan view.

5 Triangle X has vertices (2, 2), (4, 3) and (5, 2).

a Draw triangle X on a grid, with x and y from −6 to 6.

b **i** Reflect triangle X in the line $y = x$. Label the image Y.

ii Write down the coordinates of the invariant point during the reflection.

c Reflect triangle Y in the line $y = 1$. Label the image Z.

d Describe the single transformation that maps triangle X to triangle Z.

6 A ship travels 15 km from A on a bearing of 070°. At B, it travels 10 km on a bearing of 150° to reach C.

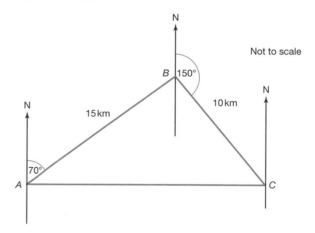

Work out

a the distance from A to C, to 3 s.f.

b the bearing of A from C to the nearest degree.

Probability

The basics of probability

The probability scale

The probability scale goes from 0 to 1. Probabilities can be expressed as decimals or fractions.

The probability formula

If all the experiments leading to an outcome are unbiased, the probability of that outcome is calculated like this:

$$\text{Probability} = \frac{\text{number of ways something can happen}}{\text{total number of possible outcomes}}$$

WORKIT!

Find the probability of throwing an even number on an unbiased dice.

$$\text{Probability} = \frac{\text{number of ways something can happen}}{\text{total number of possible outcomes}} = \frac{3}{6} = \frac{1}{2}$$

The sum of probabilities

Suppose the probability that it rains tomorrow is $\frac{5}{8}$, then the probability that it does not rain tomorrow is $\frac{3}{8}$. When only one result can happen at a time the probabilities add up to 1:

$$\text{P(event occurs)} + \text{P(event does not occur)} = 1$$

If there are three possible probabilities, A, B and C, and only one can occur at once:

$$\text{P}(A) + \text{P}(B) + \text{P}(C) = 1$$

WORKIT!

A box contains coloured balls that are red, blue or green.

The table shows the number of balls of each colour.

	Red	**Blue**	**Green**
Number of balls	$2x + 1$	7	$x + 2$

A ball is chosen at random. The probability of the ball being blue is $\frac{7}{25}$

Calculate the probability of choosing a red ball.

NAILIT!

If you throw a normal dice, the chance of it landing on a particular side is equal to the chance of it landing on any other side and the dice is said to be **unbiased** or **fair**. If there was a greater probability of the dice landing on a particular number then the dice would be **biased**.

Number of even numbers = 3 (2, 4 and 6).

Total number of numbers = 6.

Always cancel fractions, as otherwise you may lose marks.

There are only red, blue and green balls in the bag, so the total probability must add up to 1.

$$P(blue) = \frac{number\ of\ blue\ balls}{total\ number\ of\ balls} = \frac{7}{25}$$

So total number of balls = 25

$$2x + 1 + 7 + x + 2 = 25$$

Add up the number of balls to create an equation.

$$3x + 10 = 25$$

$$3x = 15$$

$$x = 5$$

Number of red balls $= 2x + 1 = 2 \times 5 + 1 = 11$

Use the calculated value of x to work out the number of red balls.

Probability of choosing a red ball $= \frac{11}{25}$

Sample space

In order to work out probability you need to be able to work out **outcomes**. The outcomes are the different possible combinations of what could happen. For example, when you toss two coins the possible outcomes are (HH, HT, TH and TT). You can use a **sample space** to show all the possible outcomes.

WORKIT!

A fair dice is rolled and a fair triangular spinner is spun. The scores are added together.

The outcomes are all the different ways of throwing the dice and spinning the spinner.

a Produce a sample space showing all the possible outcomes.

Spinner

		1	2	3
	1	2	3	4
	2	3	4	5
Dice	3	4	5	6
	4	5	6	7
	5	6	7	8
	6	7	8	9

The first column shows the score on the dice and the first row the score on the spinner.

The total scores are then filled in.

b Write down the total number of outcomes.

Total number of outcomes = 18

The number of outcomes is the number of totals in the sample space diagram.

c Find the probability of obtaining a total score of 7.

$$P(7) = \frac{number\ of\ ways\ of\ scoring\ 7}{total\ number\ of\ outcomes} = \frac{3}{18} = \frac{1}{6}$$

The dice and spinner are both fair, so all the outcomes are equally likely.

NAILIT!

An event is an outcome or group of outcomes you are interested in, such as 'it rains tomorrow' or 'it rains on Saturday and is fine on Sunday'.

You can work out the total number of outcomes using the formula:

Total number of outcomes = number of ways each action can be carried out, multiplied together

For the example above, total number of outcomes = number of dice outcomes × number of spinner outcomes: 6 × 3 = 18.

WORKIT!

A restaurant has a fixed price menu offering a choice of 4 starters, 6 mains and 5 desserts.

a How many different meals are possible? Assume that all the meals consist of one starter, one main and one dessert.

Total number of meals = 4 × 6 × 5 = 120

b What is the probability of choosing the first item on each of the starters, mains and desserts menus?

P(1st starter, 1st main, 1st dessert) = $\frac{1}{120}$

> Each outcome (combination of starter, main and dessert) is equally likely.

WORKIT!

Four fair dice are rolled.

a Calculate the number of possible outcomes.

Total number of outcomes = number of ways each action can be carried out, multiplied together

= 6 × 6 × 6 × 6 = 1296

b Calculate the number of possible ways of all the dice showing prime numbers.

Number of prime numbers on 1 dice = 3

> The possible prime numbers are 2, 3 and 5.

Number of possible outcomes for all prime numbers = 3 × 3 × 3 × 3 = 81

c Calculate the probability of all the dice showing prime numbers.

Probability of all prime numbers = $\frac{81}{1296} = \frac{1}{16}$

DOIT!

Two coins are tossed. Work out the probability of one coin landing heads and the other landing tails.

CHECKIT!

1 A five-sided spinner is spun 3 times. Find the probability of the spinner showing an even number every time.

2 There are 20 counters in a bag. 4 are black, 7 are red and 9 are green. One counter is picked at random from the bag.

a Write down the probability of obtaining a white counter.

b Calculate the probability that the counter picked is black, giving your answer as a fraction in its simplest form.

c Calculate the probability that the counter picked is **not** green.

Probability experiments

A probability experiment is an experiment, trial or observation that can be repeated numerous times under the same conditions. Each outcome of a probability experiment must not be affected by any other outcome and cannot be predicted with certainty.

Relative frequency

Relative frequency is an estimate of probability. For example, suppose you rolled a fair (i.e. unbiased) dice 500 times and recorded your results.

Relative frequency $= \dfrac{\text{frequency of event}}{\text{total frequency}}$

Score	Frequency
1	91
2	77
3	99
4	74
5	93
6	66

Total frequency (see table) $= 91 + 77 + 99 + 74 + 93 + 66 = 500$.

The relative frequency for a score of $3 = \dfrac{99}{500} = 0.198$.

The relative frequencies can be used to create a table similar to the results table.

Score	1	2	3	4	5	6
Relative frequency	$\frac{91}{500} = 0.182$	$\frac{77}{500} = 0.154$	$\frac{99}{500} = 0.198$	$\frac{74}{500} = 0.148$	$\frac{93}{500} = 0.186$	$\frac{66}{500} = 0.132$

The relative frequency can be used to give an estimate of the probability.

The theoretical probability of a particular score on an unbiased dice $= \frac{1}{6}$ or 0.1666… as a decimal.

The relative frequencies are different from this value.

The more times the experiment is conducted, the closer the relative frequencies will be to the theoretical probability $\left(\frac{1}{6}\right.$ in this case$\left.\right)$. ◄

> Note that all these relative frequencies add up to 1.

> The relative frequency will approach the theoretical frequency only over a very large number of trials.

NAIL IT!

Examples of probability experiments include:

- tossing a coin: two possible outcomes, heads or tails
- rolling a six-sided dice: six possible outcomes, the numbers 1 to 6
- selecting a numbered ball (1–50) from a bag: 50 possible outcomes.

WORKIT!

A five-sided spinner with sides numbered from 1 to 5 has the following relative frequencies of scores.

Score	1	2	3	4	5
Relative frequency	x	0.05	$2x$	0.15	0.20

Calculate the value of x.

$x + 0.05 + 2x + 0.15 + 0.20 = 1$ ◄

> The total relative frequencies have to add up to 1.

$3x + 0.4 = 1$

$3x = 0.6$

$x = 0.2$

NAIL IT!

Always check that you have taken into account all the possibilities before deciding that the probabilities add up to 1.

WORKIT!

A seed picked at random from a large batch of seeds has a probability of growing into a plant with yellow flowers of 0.12.

There are 200 seeds in a packet. Work out an estimate for how many seeds will grow into plants with yellow flowers.

Expected number
of yellow plants = probability of yellow
plant × number
of seeds

= 0.12 × 200

= 24

WORKIT!

Joshua and Jacob estimate their probabilities of being chosen to play for their school at football and rugby. Their estimates of the probabilities, which are independent, are shown in this two-way table.

	Football	Rugby
Joshua	0.7	0.6
Jacob	0.4	0.9

Calculate the probability that

a Joshua will be chosen for both sports

P(Joshua chosen for both) = 0.7 × 0.6 = 0.42

b both boys are chosen for only the rugby team.

P(Joshua not chosen for football and chosen for rugby) = 0.3 × 0.6 = 0.18

P(Jacob not chosen for football and chosen for rugby) = 0.6 × 0.9 = 0.54

P(Both boys chosen for only rugby)
= 0.18 × 0.54 = 0.0972

DOIT!

Create a revision card for relative and theoretical probability.

Expected frequency

To work out an estimate of the number of times an event occurs (**expected frequency**) we can use the following formula:

Expected frequency = probability of the event × number of events

✓ CHECKIT!

1 A six-sided spinner with numbers from 1 to 6 was spun 120 times. The results are shown in the table.

Number on spinner	1	2	3	4	5	6
Frequency	18	20	25	21	19	17

a Work out the relative frequency for a score of 3, giving your answer as a decimal to 2 decimal places.

b Work out the relative frequency for a score of 6, giving your answer as a decimal to 2 decimal places.

c Sean says that the answers to parts a and b should be the same so the spinner is not fair (i.e. it is biased). Is Sean right? Explain your answer.

2 A machine fills cans with 330 ml of cola. Out of a batch of 500 cans, 20 of them contained less than 330 ml of cola.

a Estimate the probability that the next can filled by the machine contains less than 330 ml of cola.

b The machine fills 15 000 cans per day. Estimate how many would contain less than 330 ml of cola.

3 In a batch of apples it was found that the probability of an apple being bad was $\frac{3}{40}$

In a similar batch of 600 apples, how many would be expected to be bad?

The AND and OR rules

If events A and B are **mutually exclusive**, it means that event A can happen or event B can happen but they cannot both happen at the same time.

When an event has no effect on another event, they are said to be **independent events**.

There are two important laws that need to be remembered when working out probabilities. They are the **addition rule** and the **multiplication rule**. The addition rule is sometimes called the **OR rule** and the multiplication rule is sometimes called the **AND rule**.

NAILIT!

Note the difference between independent events and mutually exclusive events.

SNAP IT!   AND and OR rules

You must learn these formulae.

Probability of A **OR** B happening: $P(A \text{ OR } B) = P(A) + P(B)$

$P(A \text{ OR } B) = P(A) + P(B) - P(A \text{ AND } B)$

Probability of A **AND** B happening: $P(A \text{ AND } B) = P(A) \times P(B)$

This formula only applies to mutually exclusive events.

This formula applies to events that can happen at the same time.

This formula only applies to independent events.

WORKIT!

1   A bag contains 3 red balls and 7 black balls. Ahmed picks one at random and notes the colour. He puts it back into the bag and picks another ball at random.

What is the probability that both balls are red?

The two events (picking each ball) are independent, as the balls in the bag are the same for each pick.

$P(\text{1st red}) = \frac{3}{10}$ and $P(\text{2nd red}) = \frac{3}{10}$

$P(\text{red AND red}) = P(\text{1st red}) \times P(\text{2nd red}) = \frac{3}{10} \times \frac{3}{10} = \frac{9}{100}$

2   There are 50 students in Year 11. 20 of them are going on a geography field trip and 25 are going on a science trip. 5 of the students are going on both trips.

   a   What is the probability that a year 11 student picked at random is going on a geography field trip?

   $P(G) = \frac{20}{50} = 0.4$

   b   Find the probability that a student picked at random is going on both trips.

   $P(G \text{ AND } S) = \frac{5}{50} = 0.1$

   c   Find the probability that a student picked at random goes on a geography or a science trip.

   $P(G \text{ OR } S) = P(G) + P(S) - P(G \text{ AND } S)$

   $= 0.4 + 0.5 - 0.1$

   $P(S) = \frac{25}{50} = 0.5$

   Use the formula

   $P(A \text{ OR } B)$
   $= P(A) + P(B) -$
   $P(A \text{ AND } B).$

   $= 0.8$

175

3 A computer manufacturer submits bids for three projects, A, B and C.

The probabilities of getting the order for projects A, B and C are $\frac{2}{5}$, $\frac{1}{3}$ and $\frac{1}{4}$, respectively. The probability of getting the order for each project is independent of the other projects.

Calculate the probability of the following events, making your methods clear. Give each answer as a fraction in its simplest form.

a The company gets all three orders.

$$P(A \text{ and } B \text{ and } C) = \frac{2}{5} \times \frac{1}{3} \times \frac{1}{4} = \frac{1}{30}$$ ← Use the AND rule.

b The company gets exactly one order.

A	Not B	Not C	$\frac{2}{5} \times \frac{2}{3} \times \frac{3}{4} = \frac{12}{60}$
Not A	B	Not C	$\frac{3}{5} \times \frac{1}{3} \times \frac{3}{4} = \frac{9}{60}$
Not A	Not B	C	$\frac{3}{5} \times \frac{2}{3} \times \frac{1}{4} = \frac{6}{60}$

Make a table showing the possible outcomes for winning exactly one project.

Use the AND rule to work out the probability of each of these outcomes.

$$P(\text{exactly one order}) = \frac{12}{60} + \frac{9}{60} + \frac{6}{60} = \frac{27}{60} = \frac{9}{20}$$

P(not getting an order) = 1 − P(getting an order)

DO IT!

Make a revision poster for the AND rule and the OR rule, including the necessary conditions.

CHECK IT!

1 James passes two sets of traffic lights on his way to work. The probability he is stopped by the first set of traffic lights is 0.2 and the probability he is stopped by the second set of lights is 0.3. The two sets of traffic lights are independent.

a Explain in relation to this question what independent events are.

b Work out the probability that James is stopped by both sets of lights.

c Work out the probability that James is not stopped by either set of lights.

2 The probability that Nala takes a packed lunch to school is $\frac{1}{4}$

The probability that Nala cycles to school is $\frac{2}{3}$

The probability that Nala has remembered her homework is $\frac{7}{8}$

a Work out the probability all of these events take place.

b Work out the probability that none of the events takes place.

Tree diagrams

The probabilities of two or more events can be shown on a **tree diagram**.

Independent events

> The two events (picking the balls out) are independent because the first ball is put back into the bag.

WORKIT!

A bag contains 5 red and 4 blue balls. A ball is picked at random from the bag, the colour is noted and the ball is put back into the bag. Another ball is picked at random and its colour noted.

a Draw a tree diagram to show these events.

b Work out the probability that both balls are red.

$$P(\text{red and red}) = \frac{5}{9} \times \frac{5}{9} = \frac{25}{81}$$

c Work out the probability that the balls are of different colours.

$$P(\text{red then blue}) = \frac{5}{9} \times \frac{4}{9} = \frac{20}{81}$$

$$P(\text{blue then red}) = \frac{4}{9} \times \frac{5}{9} = \frac{20}{81}$$

$$P(\text{different colours}) = \frac{20}{81} + \frac{20}{81} = \frac{40}{81}$$

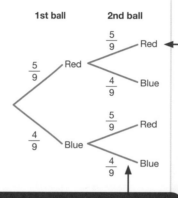

1st ball   2nd ball

> Because you can have either of these, use the OR rule.

> As the balls are replaced, the probability for the second ball is the same as for the first ball. The probabilities at each branch add up to 1 as you are certain to pick either a red or a blue ball.

> To work out the probability for a particular combination (e.g. a red and a red), multiply the probabilities on the two branches.

Dependent events

In **dependent** events, one event influences another. This tree diagram shows the same problem as above, except that this time the first ball is **not** replaced. The probability for the second ball depends on what colour was picked for the first ball.

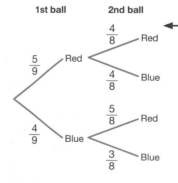

1st ball   2nd ball

> If a red ball is picked and not replaced, there are only 4 red balls and 8 balls in total for the second pick.

 NAILIT!

It is important to note that taking two things out without replacement is like taking one out without replacement and then taking the other out.

 NAILIT!

Don't cancel fractions until the end of your answer. Adding the fractions is easier if they all have the same denominator.

NAILIT!

Another way to do part c would be to find the probability that the colours of the balls are the same and then subtract this from 1.

WORKIT!

The probability that it rains on Monday is 0.3.

If it rained on Monday, the probability of it not raining on Tuesday is 0.4.

If it did not rain on Monday, the probability of it not raining on Tuesday is 0.6.

a Draw a probability tree diagram showing this information.

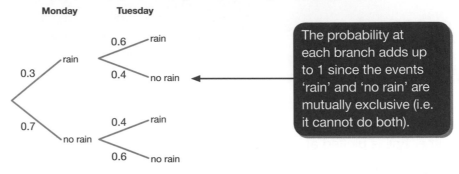

> The probability at each branch adds up to 1 since the events 'rain' and 'no rain' are mutually exclusive (i.e. it cannot do both).

b It rained on only one of the two days. Is it more likely that it rained on Monday or Tuesday? Show your working.

P(only rained on Monday) = P(rain Monday) × P(no rain Tuesday)

$$= 0.3 × 0.4 = 0.12$$

P(only rained on Tuesday) = P(no rain Monday) × P(rain Tuesday)

$$= 0.7 × 0.4 = 0.28$$

It is more likely that it rained on Tuesday.

NAILIT!

Make sure you state on which day it is more likely to have rained. It is not up to the examiner to decide this from your working out.

You may not always be told to draw the tree diagram. You may also need to use algebra to work out the answer.

> Draw a tree diagram.

WORKIT!

A box contains cartons of orange juice and apple juice in the ratio 2 : 1.
Two cartons of juice are taken out of the box at random.
The probability that both cartons are apple juice is $\frac{7}{69}$.

How many cartons of apple juice are there in the box?

Let the number of apple juice cartons be x.

> Use a variable for one part of the ratio.

$$P(2\ apple) = \frac{x}{3x} × \frac{x-1}{3x-1} = \frac{7}{69}$$

> $3x$ is total number of cartons using both parts of the ratio.

$$69x(x-1) = 7 × 3x(3x-1)$$

$$23(x-1) = 7(3x-1)$$

> Divide both sides by $3x$.

$$23x - 23 = 21x - 7$$

$$2x = 16$$

$$x = 8$$

> Solve the equation for x.

There are 8 cartons of apple juice in the box.

> If the first carton was apple, then there are $3x - 1$ cartons left, of which $x - 1$ cartons are apple.

Frequency trees

The results of a probability experiment can be displayed using a **frequency tree**. The frequency tree can then be used to estimate the probability of certain scenarios.

WORKIT!

There are 350 staff working in an office. The ratio of males to females is 4:3.

For the male staff, the ratio of the number of full-time staff to the number of part-time staff is 3:1.

For the female staff the ratio of the number of part-time staff to the number of full-time staff is 2:1.

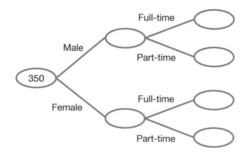

a Complete the frequency tree.

Males : females = 4 : 3   7 parts = 350, so 1 part = 50

Number of males = 4 × 50 = 200

Number of females = 3 × 50 = 150

Males: full-time : part-time = 3 : 1   4 parts = 200, so 1 part = 50

Number of males full-time = 3 × 50 = 150

Number of males part-time = 1 × 50 = 50

Females: part-time : full-time = 2 : 1   3 parts = 150, so 1 part = 50

Number of females part-time = 2 × 50 = 100

Number of females full-time = 1 × 50 = 50

> Use the ratios to work out the numbers of males and females and the numbers of full- and part-time staff.

> Fill in the frequency tree with the calculated values.

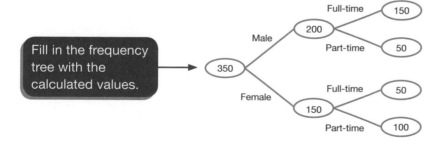

There are 150 female staff out of a total of 350 staff.

Out of 150 female staff, 50 of them work full-time.

b Using the frequency tree, work out the probability that a member of staff chosen at random is a female who works full-time.

Probability a female is chosen $= \frac{150}{350} = \frac{3}{7}$

Probability that the female works full-time $= \frac{50}{150} = \frac{1}{3}$

P(female who works full-time) $= \frac{3}{7} \times \frac{1}{3} = \frac{1}{7}$ Use the AND rule.

DO IT!

Write some revision cards containing all the main points and formulae for probability.

✓ CHECK IT!

1 There are 3 red and 7 blue counters in a bag. Two counters are picked at random, one at a time, from the bag.

Work out the probability that:

a two red counters are chosen

b a red and a blue counter are chosen.

2 Hannah puts green and blue balls into an empty bag in the ratio 2 : 3.

Zak takes at random two balls from the bag. The probability he takes two blue balls is $\frac{33}{95}$.

How many balls did Hannah put in the bag?

3 Saskia goes to the airport to catch a flight. The probability that there is a long queue to check in is 0.7. The probability that there is a long queue at security is 0.5.

a Copy and complete the tree diagram.

b Calculate the probability that there is a short queue at both check-in and security.

c Calculate the probability of a long queue at either check-in or security, or at both.

Check in

Security check

long queue

0.7

0.5 — long queue

...... — short queue

...... — long queue

...... short queue

...... — short queue

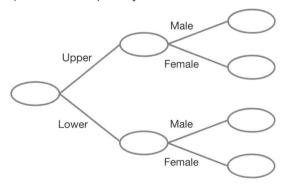

4 A secondary school has an upper school of 450 students and a lower school of 500 students. In the upper school, 60% of the students are male and in the lower school 55% are male.

a Copy and complete the frequency tree.

Male

Upper

Female

Lower

Male

Female

b A student is chosen at random from the school. Find the probability that the student is male.

Venn diagrams and probability

Venn diagrams can be used to work out probability.

The **universal set** is the whole collection of things, called **elements**, being considered. Usually an element is a number (e.g. {3, 5, 8} or an item {Jack, Queen, King}).

Sets are parts of the universal set and are shown as labelled circles/ovals on the Venn diagram. Sets contain some of the elements in the universal set and no other elements.

> The universal set is labelled with the Greek letter ξ (ξ, pronounced ksi).

The Venn diagram shows the universal set ξ = {1, 2, 3, 4, 5, 6, 7, 8, 9, 10},
set A = {2, 3, 5, 6, 7, 8} and set B = {1, 2, 4, 6, 9}.

> Curly brackets are used to show sets.

The number of elements in set A is referred to as n(A): in this example n(A) = 6.

SNAPIT! Venn diagrams

Here are the terms used for different areas on the Venn diagram:

Expression	Term	Meaning	Set of numbers
A ∩ B	A intersect B	All the elements in set A which are also in set B	{2, 6}
A ∪ B	A union B	All the elements in both set A and set B	{1, 2, 3, 4, 5, 6, 7, 8, 9}
A′	A complement	All the elements in the universal set that are not in set A	{1, 4, 9, 10}

WORKIT!

The Venn diagram shows all the factors of 36 and 40.

Set A is the set of all the factors of 36.

Set B is the set of all the factors of 40.

a Write down all the numbers in A ∩ B and describe their significance.

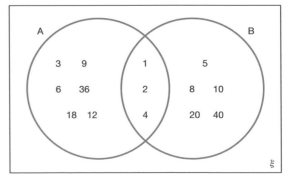

A ∩ B = {1, 2, 4}, all the numbers that are a factor of both 36 and 40

All the numbers in the intersection between A and B.

b A number is picked at random from the set A ∪ B.
Find the probability that it is an even number.

P(even number in A ∪ B) = $\frac{10}{14}$ = $\frac{5}{7}$

> A ∪ B is the set of all the numbers in A and B: 14 numbers of which 10 are even.

c A number is picked at random from the set B'.
Find the probability that the number is a multiple of 9.

B' = {3, 6, 9, 12, 18, 36}

P(multiple of 9 in B') = $\frac{3}{6}$ = $\frac{1}{2}$

> Three elements of B' are multiples of 9.

WORKIT!

There are 43 members in a youth club.
25 play badminton, 16 play chess and 20 play tennis. 5 play all three games. 2 play chess and tennis but not badminton. 7 play tennis and badminton only. 6 members play only chess. 4 members do not play any games.

> First add all the easy information to the Venn diagram.

a Draw a Venn diagram showing the above information.

16 play chess, so number who play badminton and chess only
= 16 − (6 + 5 + 2) = 3

25 play badminton, so the number who play badminton only
= 25 − (3 + 5 + 7) = 10

20 play tennis, so the number who play tennis only = 20 − (7 + 5 + 2) = 6

Use the information in the question to work out the other values for the Venn diagram.

> Put the values into the diagram as you work them out.

b A member of the club is picked at random. Find the probability that this member only plays badminton.

P(badminton only) = $\frac{10}{43}$

c Out of those members who play badminton, find the probability that they play only **one** other game.

P(badminton and only one other game)
= $\frac{7 + 3}{10 + 7 + 3 + 5}$ = $\frac{10}{25}$ = $\frac{2}{5}$

NAILIT!

Check that all the numbers on the Venn diagram add up to the total given in the question.

The conditional probability formula

Conditional probability is where the probability of one event **depends on** another event already having happened.

This means that the probability of events A **and** B occurring is the probability of event A occurring **given that** event B has occurred, multiplied by the probability that event B occurs.

SNAPIT!

If two events are **dependent** then:

P(A **AND** B) = P(A **GIVEN** B) × P(B)

P(A **AND** B) can be written P(A ∩ B).

P(A **GIVEN** B) can be written P(A|B).

STRETCHIT!

Using the more formal notation, if two events are **dependent** then:

P(A ∩ B) = P(A|B) × P(B)

Two events A and B are **independent** if, and only if:

P(A **GIVEN** B) = P(A)

WORKIT!

Jack either rides his bike or takes a bus to school. The probability that he takes a bus to school is 0.4. If he rides his bike to school, the probability that he is late is 0.1.

What is the probability that Jack rides his bike to school and is not late?

Let the probability that Jack rides his bike to school = P(B)

Let the probability that Jack is not late for school = P(A)

P(B) = 1 − 0.4 = 0.6

P(A given B) = P(Jack is not late given he rides his bike to school)
        = 1 − 0.1 = 0.9

P(A and B) = P(A given B) × P(B)

        = 0.9 × 0.6

        = 0.54

> Use the formula for conditional probability.

> Look at what you are asked to find. The two events are called A and B. Event B is riding his bike to school and event A is him not being late.

An alternative approach is to construct a tree diagram.

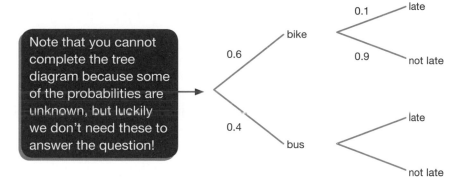

Note that you cannot complete the tree diagram because some of the probabilities are unknown, but luckily we don't need these to answer the question!

From the tree diagram

P(Jack rides his bike to school and is not late) = 0.6 × 0.9

$$= 0.54$$

CHECK**IT!**

1 The diagram below is a Venn diagram.

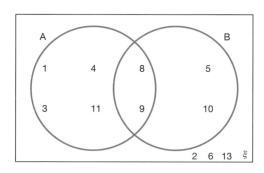

a Write down the numbers that are in the set:

   i A ∪ B   **ii** A ∩ B   **iii** A′

b One of the numbers is picked at random. Find P(B′).

2 In a sixth form of 180 students:
84 students take at least one science
40 study all three sciences
1 student studies only physics
12 study physics and biology only
48 study physics and chemistry
15 study biology and chemistry, but not physics
65 study chemistry.

a A student is chosen at random from those who take science. What is the probability that the student takes all three sciences?

b A student is chosen at random from those who take science. What is the probability that the student takes only one science?

c Out of those students studying physics, find the probability that they also study chemistry.

3 The universal set is the set of integers from 1 to 20. Set A is the set of square numbers. Set B is the set of all even numbers.

a Complete the following Venn diagram.

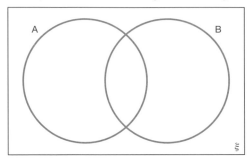

b An integer is chosen at random. Find the probability that it is not an even or square number.

4 $P(A) = \frac{1}{10}$, $P(B) = \frac{1}{4}$ and $P(A \cap B) = \frac{1}{48}$

Find

a $P(A|B)$

b $P(B|A)$.

5 An insurance company records details of claims made during cold or warm weather.

	Cold weather	Warm weather
Burst pipes	165	74
Not burst pipes	235	236

a How many claims were made during the period represented by the table?

b Let A be the event 'cold weather' and B the event 'burst pipes'.

Are A and B independent? You must show your workings.

6 The Venn diagram shows the probabilities attached to two events P and Q.

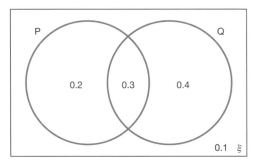

Find the probabilities of

a event P given event Q has occurred

b event Q given event P has occurred.

Probability

1 A bag contains 4 red counters, 3 green counters and 3 white counters.

A counter is taken at random from the bag and not replaced. Then a second counter is taken from the bag.

a Complete the tree diagram.

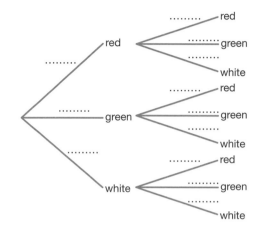

b Calculate the probability that the two counters taken from the bag are the same colour.

c Calculate the probability that the two counters taken from the bag are different colours.

2 A bag contains 3 red balls and x black balls.

Two balls are removed at random from the bag.

The probability that two black balls are chosen is $\frac{7}{15}$.

a Show that $4x^2 - 25x - 21 = 0$.

b Find the total number of balls in the bag.

c Find the probability of the two balls being different colours.

3 60 students were asked which sports they liked watching from football, tennis and motor racing.

All 60 students liked watching at least one sport.

10 students liked watching all three sports.

3 liked watching football and motor racing only.

15 liked watching tennis and motor racing.

29 liked watching motor racing.

18 liked watching football and tennis.

41 liked watching football.

a What is the probability that a student chosen at random from the group only liked to watch motor racing?

b A student is chosen from those who liked motor racing. What is the probability that the student also liked tennis?

4 In a certain country, out of 500 defendants being tried by the law courts 75% committed the crime. Of those who committed the crime, 80% were found guilty. Of those who did not commit the crime, 20% were found guilty.

a Complete the frequency tree to show this information.

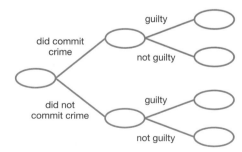

b A defendant was chosen at random from the 500 defendants. Find the probability that the defendant was found guilty.

c A defendant was chosen at random from those who did not commit the crime. Find the probability that the defendant was found guilty.

Statistics

Sampling

SNAPIT! Types of data

Discrete data: data that can only take certain values (e.g. number of pets, shoe size, hair colour).
Continuous data: measured data that can have any value (e.g. length, weight, time).
Primary data: data you collect yourself (e.g. by conducting a survey).
Secondary data: data collected by someone else (e.g. a research organisation).

It is not usually possible to carry out a survey or a poll on an entire **population** (e.g. people eligible to vote in an election, students in a school, people who attend a sporting event). Instead a smaller number of people called a **sample** are surveyed.

The sample should be large enough to represent the population (e.g. a sample of 50 out of a population of 10 000 would be too small). The larger the sample size, the more accurate the results. The sample must also be **representative** of the population (e.g. same proportion of males/females and age groups).

Random sampling

The sample should be taken randomly, with each member of the population having an equal chance of being chosen. To ensure a **random** sample, you could:

* number each person and then use the random select on a calculator or computer

* write names on paper, fold them, jumble them up in a container and then pick them out.

Bias

A **biased** sample will not reflect the population accurately. The following should be considered when sampling:

* **The time the survey is done**: doing a survey at 8.30 am outside a train station on a weekday is less likely to include people who do not work.

* **Where the survey is done**: trying to find out the number of holidays people take by interviewing people as they leave a travel agent shop would bias the results towards people who take holidays.

* **The gender of people in the survey**: the sample should have the same percentage of males/females as the population.

* **The age of the people in the survey**: the sample should have the same age composition as the population.

WORKIT!

A company has 208 employees. The table shows numbers of employees in each department.

A survey is to be conducted with a sample of 50 employees.

Sales	Administration	Manufacturing	Delivery	Accounts	Management
56	33	76	21	12	10

a Explain the difference between a population and a sample, illustrating your answer with the information above.

The population is the whole group, in this case all the employees. The sample is a smaller group chosen from the population. In this case, it is the 50 employees to be used for the survey.

b Calculate the number of employees in the sales department that should be chosen.

% of the employees who work in sales = $\frac{\text{number who work in sales}}{\text{total of all employees}} \times 100$

Find this percentage of the number in the sample.

$= \frac{56}{208} \times 100$

$= 26.9\%$

Number of sales employees in sample = 26.9% of 50 = 0.269 × 50 = 13.45

The sample will include 13 members of the sales department.

Round the calculated value to the nearest whole number.

NAILIT!

An even quicker way of doing this would be to multiply 50 by the fraction $\frac{56}{208}$

DOIT!

Here are some trusted sources of secondary data. Look at the kind of data they hold.

www.gov.uk/government/statistics

www.cia.gov/library/publications/the-world-factbook

CHECKIT!

1 A youth club has 350 members as shown in the table. The club organiser wants to survey the members to check the club is providing the sorts of activities young people like. He decides to conduct a survey of 50 members selected at random.

The table shows the number of male and female members.

Male	Female
185	165

Work out the number of males that should be chosen for the sample.

Two-way tables and pie charts

Two-way tables

Two-way tables are used to show information in two categories. The following example shows data on gender and favourite food.

WORKIT!

The two-way table gives information about the favourite food, chosen from Indian, Chinese and Italian, of 261 students in a school.

a Complete the two-way table.

	Indian	Chinese	Italian	Total
Boys	75	28	44	147
Girls	50	38	26	114
Total	125	66	70	261

> The numbers must add up across the rows and down the columns to give the totals.
>
> Start with a column or row that is only missing one value: boys liking Chinese = 147 − 75 − 44 = 28.

b A student is picked at random from the school. What is the probability that the student's favourite food is Chinese? Give your answer as a fraction in its simplest form.

$P(\text{Chinese}) = \frac{66}{261} = \frac{22}{87}$

c A boy is picked at random. Find the probability that his favourite food is Indian or Italian.

$P(\text{boy's favourite food is Indian or Italian}) = \frac{75 + 44}{147} = \frac{119}{147} = \frac{17}{21}$

Tabulating discrete and continuous data

Numerical data can be **discrete** or **continuous**. Continuous data can take any value, for example, mass and height. Discrete data can only have specific values, for example, the number of people in a queue.

Suppose you did a survey of the number of first cousins a group of 40 people had. You could list all the responses, but this would be cumbersome. Instead, put the data into **classes** (groups) in a table.

First cousins	Frequency
0–2	5
3–5	11
6–8	12
9–11	9
12 or more	3

Height of dog (h cm)	Frequency
$0 < h \leq 20$	6
$20 < h \leq 40$	25
$40 < h \leq 60$	28
$60 < h \leq 80$	10
$80 < h \leq 100$	6

With continuous data, the classes are shown using inequalities.

> Here is an example of continuous data: the heights of 75 dogs taking part in a dog show.

> $80 < h \leq 100$ is a class. There are five classes in this table.

Pie charts

A **pie chart** shows the division of a total into its separate parts. To work out the angle for each part:

1 Find the total frequency.

2 Divide 360° by the total frequency to find the number of degrees represented by one item.

3 Multiply this by each of the frequencies to find the angle for each category.

4 Use these angles to draw the pie chart.

NAILIT!

Choose the class interval such that there are 4–6 classes in the table. Always make sure that classes do not have values that overlap. For example, with two classes $0 < h \leq 20$ and $20 \leq h \leq 40$, a height of 20 cm would be in both classes.

WORKIT!

The table gives information about the manufacturers of cars in a car park.

Manufacturer	Frequency
Ford	12
Vauxhall	18
Nissan	5
Toyota	10
Mercedes	15

Draw an accurate pie chart showing this information.

$12 + 18 + 5 + 10 + 15 = 60$ ← Add up the frequencies.

60 cars $= 360°$ so angle representing 1 car $= \dfrac{360}{60} = 6°$ ← Multiply the frequencies by this angle to find the angle for each manufacturer.

Manufacturer	Frequency	Angle
Ford	12	$12 \times 6 = 72°$
Vauxhall	18	$18 \times 6 = 108°$
Nissan	5	$5 \times 6 = 30°$
Toyota	10	$10 \times 6 = 60°$
Mercedes	15	$15 \times 6 = 90°$

Always check that the angles add up to 360°.

Use compasses, a ruler and a protractor to draw the pie chart.

CHECKIT!

1 The table gives information about newt species in a pond.

Species	Frequency
Smooth newt	13
Great crested newt	7
Palmate newt	10

a Draw a pie chart to show this information.

b On a pie chart for another pond the angle sector for the smooth newt was 135°. Does this mean that fewer smooth newts were in the other pond? Explain your answer.

Line graphs for time series data

Some quantities (e.g. sales of fireworks or barbecues, temperature, rainfall) vary in a repeating pattern over time. When these quantities are plotted against time they show repeating peaks and troughs, making it difficult to spot **trends** (e.g. increasing or decreasing values over time).

Using a **moving average** makes it easier to spot trends. This smooths out the peaks and troughs so that the graph is straighter.

Showing a time series graphically

Suppose you have the following data.

	Q1 2015	Q2 2015	Q3 2015	Q4 2015	Q1 2016	Q2 2016	Q3 2016	Q4 2016
Sales (thousands)	7	12	6.5	4	6	11	5.5	3.5

This time series is plotted with the time on the x-axis and the sales on the y-axis.

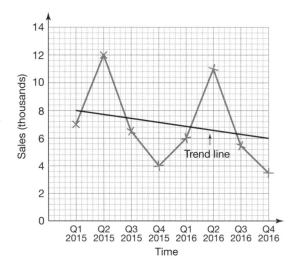

You can see the seasonal ups and downs from the graph. What we really need to look at is the general trend. If you look at the two peaks you can see that they are getting lower. The troughs are getting slightly deeper.

You can now draw a trend line which runs between the peaks and troughs showing the downward trend. This line can be extended beyond the times for which you have data, to predict future sales.

The line shows that there is a slight general downward trend in sales. However, it should be noted that this is only an approximate prediction.

STRETCHIT!

The graph on the previous page shows a lot of variation. A better way to see the trend would be to find the 4-point moving average.

	Q1 2015	Q2 2015	Q3 2015	Q4 2015	Q1 2016	Q2 2016	Q3 2016	Q4 2016
Sales (thousands)	7	12	6.5	4	6	11	5.5	3.5

To find the 4-point moving average (to 1 decimal place):

1. Find the mean of the first four values:

$$\frac{7 + 12 + 6.5 + 4}{4} = 7.4$$

Plot this value at the centre of the period covered by the four values (i.e. midway between Q2 2015 and Q3 2015).

2. Move one unit of time along (in this case, Q2 2015) and find the mean of the next four values:

$$\frac{12 + 6.5 + 4 + 6}{4} = 7.1 \text{ plotted midway between Q3 and Q4 2015}$$

3. Repeat step 2 until the end of the data:

$$\frac{6.5 + 4 + 6 + 11}{4} = 6.9 \text{ plotted midway between Q4 2015 and Q1 2016}$$

$$\frac{4 + 6 + 11 + 5.5}{4} = 6.6 \text{ plotted midway between Q1 and Q2 2016}$$

$$\frac{6 + 11 + 5.5 + 3.5}{4} = 6.5 \text{ plotted midway between Q2 and Q3 2016}$$

Plotting this 4-point moving average on the original graph would give a more accurate trend line than the simple straight line.

CHECKIT!

1 The sales of a maths textbook for the first 6 months of the year are recorded here.

Month	Jan	Feb	Mar	Apr	May	June
Sales	936	939	903	870	882	810

Describe the trend in sales over the period for the data.

2 This table records the percentage of people who live in a village visiting their local shop.

Year	1996	2000	2004	2008	2012	2016
Percentage	70	65	57	53	51	49

a Plot the time series graph. Make the x-axis go up to at least 2025.

b Describe the trend in the percentage of people using the local shop over the last 25 years.

c If the trend continues, what percentage of the people are likely to be using the village shop in 2025?

d Explain why your prediction in part c may not be reliable.

Averages and spread

In statistics there are three **averages**: mean, mode and median.

The **mean** of a set of n items is the sum of the items divided by n:

$$\text{Mean} = \frac{\text{total of the items}}{\text{number of items }(n)}$$

WORKIT!

The mean age of four members in a group is 22. One new member joins the group and the mean age falls to 21. How old is the new member?

Mean age of four members $= 22 = \dfrac{\text{total age}}{4}$

Total age $= 22 \times 4 = 88$

Let the age of the member joining the group $= x$

Mean age of 5 members $= \dfrac{88 + x}{5}$

$21 = \dfrac{88 + x}{5}$

$105 = 88 + x$

$x = 17 \text{ years}$

NAILIT!

Use the same method to estimate the mean from a graph by creating a table of the data.

The **median** is the halfway value.

To work out the median of a set of numbers:

1. Arrange the numbers in ascending order (i.e. smallest first).
2. Find the middle number; if there are two middle numbers, find the mean of them.

The **mode** is the most common value (the value with the greatest frequency).

To find the mode:

1. List the numbers in ascending order.
2. Identify the value that occurs most often. This is the mode.

Sometimes there may be two or more modes.

You may also need to work out the lower and upper **quartiles**. The **lower quartile** is the boundary between the first and second quarters of data. The **upper quartile** is the boundary between the third and fourth quarters of data.

NAILIT!

If you are asked to work out the mode from a histogram remember that the modal class will not always be the highest bar – work out the area of each bar to find the frequency.

WORKIT!

Here is a set of 14 values:

7, −8, 0, 4, −1, −2, 3, 2, 3, 2, −3, 2, 1, 1

Find the mean, median and mode.

Mean = $\dfrac{7 + (-8) + 0 + 4 + (-1) + (-2) + 3 + 2 + 3 + 2 + (-3) + 2 + 1 + 1}{14}$

$= \dfrac{11}{14}$

$= 0.79$ (to 2 d.p.)

Arranging the values in ascending order:

−8, −3, −2, −1, 0, 1, 1, 2, 2, 2, 3, 3, 4, 7

Median = $\dfrac{1 + 2}{2} = 1.5$ ◄ — There is an even number of values: the middle values are the 7th and 8th ones.

Mode = 2 ◄ — 2 is the most frequent value in the set, as it appears three times.

NAILIT!

Always check the number of values after rearranging them to make sure you have included all of them.

NAILIT!

It is easier to spot the mode after putting the numbers in order.

To find the **lower quartile** of a set of numbers:

1 Arrange the data in ascending order.

2 Split the data into two equal groups of items. If there is an odd number of items, include the middle one.

3 Find the median of the **lower** half of the data set.

To find the **upper quartile** of a set of numbers, use the same process, but find the median of the **upper** half of the numbers in the set.

WORKIT!

Find the lower quartile and the upper quartile for the following shoe sizes of 20 Year 11 boys.

7 10 11 8 8 7 6 8 9 9 8 9 10 10 8 6 8 9 10 7

6 6 7 7 7 8 8 8 8 8 8 9 9 9 9 10 10 10 10 11 ◄ — To find the lower quartile first write the shoe sizes in order of size.

Lower half of data is:

6 6 7 7 ⑦ ⑧ 8 8 8 8 ◄ — Now take the lower half of the data and find the median.

Median of this set of data = $\dfrac{(7 + 8)}{2} = 7.5$

Hence lower quartile of complete set of data = 7.5

Upper half of data is:

8 9 9 9 ⑨ ⑩ 10 10 10 11 ◄ — Now take the upper half of the data and find the median.

Median of this set of data = $\dfrac{(9 + 10)}{2} = 9.5$

Hence upper quartile of complete set of data = 9.5

DOIT!

Write a revision card summarising how to find each average and the lower and upper quartiles.

For more on lower and upper quartiles see cumulative frequency graphs (pages 199–200).

Estimating the mean of a grouped distribution

When there is a large number of data items, it is usual to put them into classes (e.g. 20 < *t* ≤ 30). You can't work out an exact mean, as you don't have the individual data values. However, you can **estimate** the mean using the mid-interval value for each class.

WORKIT!

Sally recorded the times it took her to walk to school over 30 days. The table gives information about her data.

Time (*t* minutes)	Frequency		
20 < *t* ≤ 30	4		
30 < *t* ≤ 40	8		
40 < *t* ≤ 50	16		
50 < *t* ≤ 60	2		

> **NAIL**IT!
> The blank columns in this table indicate that there is some working out to do.

Work out Sally's mean time to walk to school in minutes, correct to 3 significant figures.

Time (*t* minutes)	Frequency	Mid-interval value	Frequency × mid-interval value
20 < *t* ≤ 30	4	25	4 × 25 = 100
30 < *t* ≤ 40	8	35	8 × 35 = 280
40 < *t* ≤ 50	16	45	16 × 45 = 720
50 < *t* ≤ 60	2	55	2 × 55 = 110

> Create two new columns: the mid-interval value and frequency × mid-interval value. The mid-interval value is the midpoint of the range.

$$\text{Estimated mean} = \frac{100 + 280 + 720 + 110}{30} = 40.33 = 40.3 \text{ minutes (to 3 s.f.)}$$

> Divide the total of all the frequency × mid-interval values by the total frequency.

✓ CHECKIT!

1 The mean age of boys in a youth club is 13 and the mean age of the girls is 14. There are 10 boys and 12 girls. Calculate the mean age for the club members.

2 There are 25 students in a maths class. The 10 boys in the class got a mean mark of 50% in a maths exam. The 15 girls got a mean mark of 62% in the same exam.

Joshua says that the mean mark for the entire class can be worked out as $\frac{50 + 62}{2} = 56\%$.

Explain why Joshua is wrong and work out the correct mean mark, showing your working.

3 A company keeps a record of its employees' taxi expenses over a month.

Cost (£*C*)	Frequency	Mid-interval value	Frequency × mid-interval value
0 < *C* ≤ 4	12		
4 < *C* ≤ 8	8		
8 < *C* ≤ 12	10		
12 < *C* ≤ 16	5		
16 < *C* ≤ 20	2		

a Copy and complete the table.

b Calculate an estimate for the mean taxi fare for the month.

Histograms

Histograms consist of vertical bars, but they are different from bar charts. In a histogram it is the **areas** of the bars that represent the frequency, not the height.

Histograms:

- have numbers on both axes
- can have bars of different widths
- have frequency density rather than frequency on the vertical axis.

DOIT!

Make an annotated revision card showing the key features of a histogram.

SNAPIT! Histograms

Frequency (area of bar) = **class width** (width of bar) × **frequency density** (the bar's height)

You often need to rearrange this formula to give:

Frequency density = $\dfrac{\text{frequency}}{\text{class width}}$

NAILIT!

You must remember this formula.

The class width is found by subtracting the higher value from the lower value of the class interval: for $5 < d \leq 9$, class width = $9 - 5 = 4$.

WORKIT!

1 The table gives information about the diameter of ripe tomatoes in a greenhouse.

Diameter (d centimetres)	Frequency
$0 < d \leq 2$	14
$2 < d \leq 3$	16
$3 < d \leq 5$	24
$5 < d \leq 9$	30

Draw a histogram to show this information.

Diameter (d centimetres)	Frequency	Frequency density
$0 < d \leq 2$	14	$\frac{14}{2} = 7$
$2 < d \leq 3$	16	$\frac{16}{1} = 16$
$3 < d \leq 5$	24	$\frac{24}{2} = 12$
$5 < d \leq 9$	30	$\frac{30}{4} = 7.5$

Draw the histogram with diameter (d cm) on the horizontal axis and frequency density on the vertical axis.

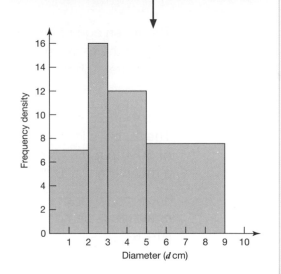

Work out the frequency densities by dividing each frequency by its class width.

2 This histogram shows the ages of football team supporters travelling on a plane to a match.

Notice that there are no units for frequency density on the vertical axis. You will need to work these out.

Put a scale on the vertical axis using the frequency density for the class $20 \le a < 35$.

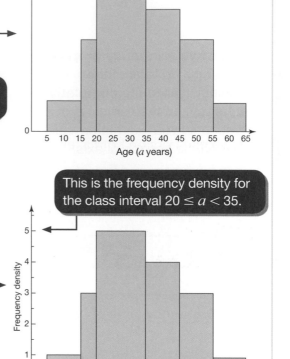

There are 75 supporters with ages in the range $20 \le a < 35$.

This is the frequency density for the class interval $20 \le a < 35$.

Work out the number of supporters with ages in the range $35 \le a < 45$.

For $20 \le a < 35$, frequency density

$= \dfrac{\text{frequency}}{\text{class width}} = \dfrac{75}{15} = 5$

Work out the frequency density for the class $20 \le a < 35$.

Read this off the graph.

For $35 \le a < 45$,

Frequency = class width × frequency density = $10 \times 4 = 40$

The number of supporters with ages in the range $35 \le a < 45$ is 40.

STRETCH IT!

To estimate the median from a histogram, first work out the total area of the bars, halve it, and then work from the left to find out which bar this takes you to. In the above example, half the area is 18, which falls in class $20 \le a < 35$. Now find how far along this bar you need to be. There are 5 values below it and 16 above it, so work out $18 - 5 = 13$ and $18 - 16 = 2$, and use these as a ratio: 13 : 2 to calculate the distance along the bar as $\dfrac{13}{13 + 2} \times (35 - 20) = \dfrac{13}{15} \times 15 = 13$. So estimate of median is $20 + 13 = 33$.

CHECK IT!

1 The table and histogram show information about the masses of kittens.

Copy and complete the table and the histogram.

Mass (*m* pounds)	Frequency
$0.0 < m \le 0.5$	30
$0.5 < m \le 1.0$	
$1.0 < m \le 1.5$	13
$1.5 < m \le 2.0$	21
$2.0 < m \le 2.5$	
$2.5 < m \le 3.0$	38

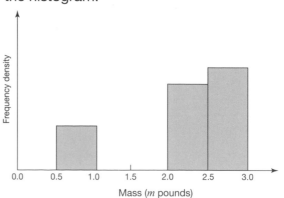

Cumulative frequency graphs

Cumulative frequency graphs have a running total of the frequency (called the **cumulative frequency**) on the vertical axis and the quantity you want information about on the horizontal axis.

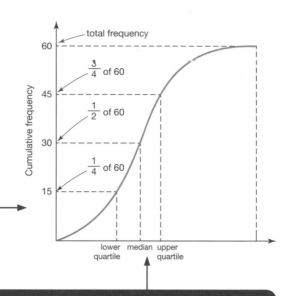

You can find out the median and the lower and upper quartiles by reading them off the graph.

Interquartile range = upper quartile − lower quartile

NAILIT!

The x-axis shows all the possible values in your sample, evenly spaced. The y-axis reading shows how many items of data are at, or below, the corresponding value on the x-axis. When you get to the total frequency (60 in this example), all the data are at, or below, the corresponding value on the x-axis.

WORKIT!

A perfume counter in a duty free store keeps records of the prices of bottles of perfume they sell in a day.

Cost (£ C)	Frequency	Cumulative frequency
$0 < C \leq 20$	28	
$20 < C \leq 40$	42	
$40 < C \leq 60$	35	
$60 < C \leq 80$	45	
$80 < C \leq 100$	50	
$100 < C \leq 120$	15	

a Write down the modal class interval.

Modal class interval is $80 < C \leq 100$.

This is the class interval with the highest frequency.

b Copy and complete the cumulative frequency table.

Cost (£ C)	Frequency	Cumulative frequency
$0 < C \leq 20$	28	28
$20 < C \leq 40$	42	70
$40 < C \leq 60$	35	105
$60 < C \leq 80$	45	150
$80 < C \leq 100$	50	200
$100 < C \leq 120$	15	215

$28 + 42 = 70$

The last entry in cumulative frequency is the total frequency. In this case it is the total number of bottles of perfume sold.

199

c Draw the cumulative frequency diagram.

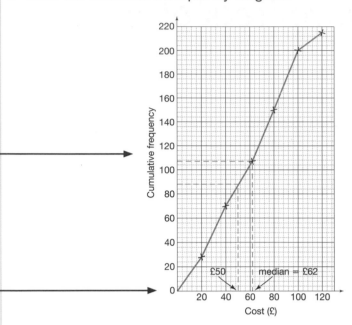

Plot each cumulative frequency value at the upper class boundary, i.e. 20, 40, 60…

Plot a value of zero at the lower boundary of the first class interval.

Draw a line from cumulative frequency $= \frac{215}{2} = 107.5$ to the graph line and read off the corresponding value from the x-axis.

d Use the graph to estimate the median price for the bottles of perfume sold that week.

Median cost = £62

e Use the graph to estimate the number of bottles of perfume sold that week that cost over £50.

Cumulative frequency for a cost of £50 is 88.

Number sold costing over £50 = 215 − 88 = 127

Look at 50 on the cost axis and read the cumulative frequency from the other axis.

DOIT!

Draw a flowchart for drawing a cumulative frequency graph.

CHECKIT!

1 The cumulative frequency table shows the waiting times in minutes for passengers to pass through security at an airport.

Time (t minutes)	Cumulative frequency
$0 < t \le 5$	15
$0 < t \le 10$	50
$0 < t \le 15$	122
$0 < t \le 20$	145
$0 < t \le 25$	165
$0 < t \le 30$	175
$0 < t \le 35$	180

a Plot a cumulative frequency graph for the information.

b Use your graph to find the median waiting time.

c Find

   i the upper quartile

   ii the lower quartile

   iii the interquartile range.

d People who wait more than 27 minutes are often in danger of missing their flight. Work out what percentage of passengers waited 27 minutes or more at security. Give your answer to 1 decimal place.

Comparing sets of data

Box plots

Box plots are diagrams used to summarise important values in a set of data. They show:

- lower quartile
- median
- upper quartile
- lowest value
- highest value.

To measure the spread of a data set using a box plot, you can use:

- **range** = highest value − lowest value

- **interquartile range** (IQR) = upper quartile − lower quartile.

The interquartile range measures the spread of the middle half of the data. As it is the middle half of the data it will not be affected by extremely large or small values (**outliers**).

> Box plots give no information about individual values or how many values there are.

WORKIT!

The number of speeding fines issued in a city per week was recorded over a one-year period:

- lower quartile = 40
- median = 55
- upper quartile = 65
- lowest number = 30
- highest number = 80.

a Draw a box plot to summarise the data.

b Work out

i the range

Range = highest value − lowest value
= 80 − 30 = 50

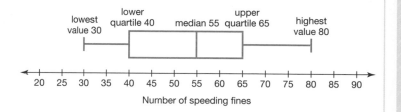

ii the interquartile range.

Interquartile range = upper quartile − lower quartile
= 65 − 40 = 25

Comparing data using box plots

Two sets of data can be compared using box plots. For example, these box plots show the results for a mock GCSE maths exam taken by two groups of students, A and B.

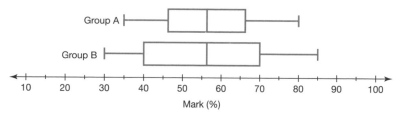

The box plots can be used to compare the two sets of data:

- Both groups have the same median mark of 56%.

- Group A has an upper quartile of 66% and a lower quartile of 46%, giving an interquartile range of 20%.

- Group B has an upper quartile of 70% and a lower quartile of 40% giving an interquartile range of 30%.

- The lower interquartile range of group A means that the marks were less spread and the middle half of the data was nearer to the median. This indicates that group A's results were more consistent.

- The range for group A is 80 − 35 = 45% and the range for group B is 85 − 30 = 55%. Therefore there was a greater variation in marks for group B.

Taking all these into account, there are reasons for arguing that group A's exam results were better than group B's.

DO IT!

Make a poster to show how two cumulative frequency graphs or box plots can be used to compare two sets of data.

Comparing data using cumulative frequency graphs and box plots

To compare sets of data you may be given two cumulative frequency graphs or a cumulative frequency graph for one set of data and a box plot for the other. From both, you can find the median, range, upper and lower quartiles and the interquartile range, and then compare the data sets.

✓ CHECK IT!

1 The box plot and table give information about the distribution of heights of sunflowers in a garden.

Height (cm)

	Height (cm)
Lowest height	38
Lower quartile	
Median	
Upper quartile	
Highest height	69

Copy and complete the table and the box plot.

2 The cumulative frequency graphs show information about the marks 100 boys and 100 girls obtained in a maths exam.

a For the boys, find
  i the range of the marks
  ii the median mark
  iii the upper quartile mark
  iv the lower quartile mark
  v the interquartile range of the marks.

b Using the median and the interquartile range, compare the distribution of the marks obtained by the boys with the marks obtained by the girls.

Scatter graphs

Types of correlation

Correlation is the relationship between variables on a scatter graph.

- **positive** correlation: the points slope upwards from bottom left to top right – as one quantity goes up, so does the other

- **negative** correlation: the points slope downwards from top left to bottom right – as one quantity goes up, the other goes down

- **no** correlation: the points do not slope in any definite direction – one quantity does not depend on the other.

Depending on how close the points are to making a straight line you can have:

- **strong** correlation, where the points are near to being in a straight line

- **weak** correlation, where the points are less closely grouped to a straight line.

Strong positive correlation

Weak negative correlation

No correlation

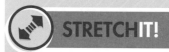

STRETCH IT!

Correlation between two quantities does not necessarily mean that one is caused by the other. For example, the increase in sales of umbrellas could be because there is a half-price sale and not because of wet weather.

Lines of best fit

The **line of best fit** is a line drawn between the points such that the total distance to the points either side of the line is approximately the same.

Sometimes a point is way out from the others. This is called an **outlier**. It represents an unusual value, or, in some cases, a mistake in the data. Ignore outliers when drawing your line of best fit.

When using a line of best fit to make estimates, extending the line past the last point could mean that the results are unreliable. You will be assuming that the trend shown by the line continues, which may or may not be the case.

WORKIT!

A supermarket sells wellington boots. Each day the shop assistant measures the amount of rainfall (cm) and the number of pairs of wellington boots sold, as shown on the scatter graph.

On a particular day there were 5.5 mm of rainfall and 4 pairs of wellington boots were sold.

a On the scatter graph, circle the point representing the data for this day.

b Describe the relationship between the amount of rainfall and the number of pairs of wellington boots sold.

Strong positive correlation: as the rainfall goes up so does the number of wellington boots sold.

c One day had 3 mm of rainfall. Estimate the number of pairs of wellington boots sold.

3 ← Draw in the line of best fit. Read the information off your line, at the 3 mm value on the horizontal axis.

DOIT!

Draw a poster showing different types of correlation.

CHECKIT!

1 The table shows the daily number of visitors to a beach and the temperature at noon.

Temperature (°C)	5	10	15	20	25	30	35
Daily visitors	17	30	40	53	60	85	90

a Draw a scatter graph for the data, with temperature on the horizontal axis.

b Draw the line of best fit.

c Use your line of best fit to estimate the number of visitors to the beach when the temperature is 27 °C.

d **i** Use your line of best fit to estimate the number of visitors to the beach when the temperature is 45 °C.

ii Explain why this estimate of the number of visitors is likely to be unreliable.

1 The table shows information about the amount of pocket money per week each student in a class receives.

Amount of pocket money (£a per week)	Number of students
$2 < a \leq 4$	6
$4 < a \leq 6$	15
$6 < a \leq 8$	9
$8 < a \leq 10$	6

a How many students are there in the class?

b Work out an estimate for the mean amount of pocket money for the class.

c Write down the modal class interval.

2 The table shows the marks 100 students got in an exam.

Mark (m)	Frequency
$0 < m \leq 20$	6
$20 < m \leq 40$	19
$40 < m \leq 60$	39
$60 < m \leq 80$	21
$80 < m \leq 100$	15

a Draw the cumulative frequency curve for this information.

b Use your graph to work out the median mark for the class.

It has been decided to set the pass mark so that only 75% of the students pass the exam.

c Use the graph to find the minimum mark needed to pass the exam.

3 Four schools, A, B, C and D, each have 150 students. They take the same examination, with a maximum mark of 80. The cumulative frequency graphs for each school are shown.

A

B

C

D
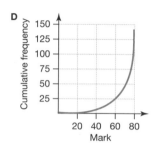

a Which school had the highest median mark?

b Which school had the largest interquartile range?

c Which school performed best in the exam? Give reasons for your answer.

4 Alex and Nadia are organising a street party for 320 people. They interview a sample of 40 people in the street.

a How should they select their sample?

b One question they asked was about what people would drink. The table shows their results.

Drink	Number of people
Cola	7
Lemonade	15
Squash	5
Tea	10
Coffee	3

How many people should they buy lemonade for?

Answers

For full worked solutions, visit:
www.scholastic.co.uk/gcse

Number

Integers, decimals and symbols

1 a 200.1   b 2.001   c 2.3   d 87
2 a 140.94   b 1.4094   c 290   d 4.86
3 −0.5, 0, 0.012, 0.12, 12
4 a $\frac{5}{0.5} = 10$ b $1\frac{5}{9} > \frac{4}{3}$ c $-3 < -1$
5 a 5   c −15   e −4
   b −8   d 6

Addition, subtraction, multiplication and division

1 a 1561   c 69.93
   b 3047   d 23.923
2 a 2819   c 8.185
   b 287   d 5.401
3 a 29798   b 29.26   c 40.768
4 a 46   b 343   c 35.4

Using fractions

1 a $\frac{16}{5} = 3\frac{1}{5}$ c $\frac{5}{8}, \frac{3}{4}, \frac{9}{10}, 1\frac{1}{5}, \frac{16}{5}$
   b $1\frac{1}{5} = \frac{6}{5}$ d $4\frac{2}{5}$ e $2\frac{23}{40}$
2 $\frac{15}{45}, \frac{4}{12}, \frac{16}{48}$
3 a $7\frac{1}{3}$ b $3\frac{1}{2}$
4 $\frac{13}{60}$

Different types of number

1 a 16   b 5   c 16
2 $2 \times 2 \times 3 \times 5 \times 5$
3 every 144 days
4 a $2^2 \times 3^3 \times 7$ b 36

Listing strategies

1 12   2 180

The order of operations in calculations

1 a 18   b 13   c 25
2 a 10   b 23   c ±5

Indices

1 a 7^{10} b 3^{-6} c 5^{20}
2 a 5^7 c $2^{10} \times 5^{-3}$
   b 6^{-3} d $7^{10} \times 11^{-1}$
3 a 1   c 16   e 18
   b 10   d $\frac{1}{5}$ (or 0.2)
4 $x = 1$

Surds

1 a $\sqrt{6}$ b 5   c 18   d 20
2 $a = 2$
3 a $3\sqrt{5}$ b $6\sqrt{2}$
4 a $\frac{16}{3\sqrt{2}} = \frac{16\sqrt{2}}{3\sqrt{2}\sqrt{2}} = \frac{16\sqrt{2}}{3 \times 2} = \frac{8\sqrt{2}}{3}$ b $8 - 2\sqrt{7}$
5 a −4   b $7 + 4\sqrt{3}$ c $5 + 3\sqrt{3}$

Standard form

1 a 0.005   b 565000
2 a 2.5×10^4 c 5×10^2
   b 1.25×10^{-3} d 1.4×10^{-2}
3 a 9×10^{-4} c 2×10^2
   b 2.4×10^3 d 8.04×10^4
4 a 1.33×10^{10} pounds
   b 26600000 people
5 a 1.55×10^4 c 5×10^2
   b 655000   d 4×10^3

Converting between fractions and decimals

1 a 0.43   b 0.375   c 0.55
2 a $\frac{4}{5}$ b $\frac{9}{20}$ c $\frac{73}{125}$
3 a $\frac{7}{9}$ b $\frac{2}{45}$ c $\frac{21}{22}$
4 a $\frac{14}{27}$ b $\frac{19}{25}$
5   Let $x = 0.7777777\ldots$
     $10x = 7.7777777\ldots$
   $10x - x = 7$
      $9x = 7$
       $x = \frac{7}{9}$
Therefore, $0.77777\ldots + \frac{2}{9} = \frac{7}{9} + \frac{2}{9} = 1$
6 $-0.9, \frac{7}{10}, \frac{4}{5}, 0.85, \frac{7}{8}$

Converting between fractions and percentages

1 a $\frac{1}{4}$ b $\frac{17}{20}$ c $\frac{17}{25}$
2 maths: 81.25%
   Charlie did better at maths.
3 a 30%   b 16%   c 42.9%

Fractions and percentages as operators

1 a £480   b £4.50   c 156kg
2 School A: 336, School B: 455

Standard measurement units

1 a 9700g   b 0.85 litres   c 205000cm
2 8.64×10^4 seconds
3 £81.60

Rounding numbers

1. **a** 1260   **c** 0.000308   **e** 1.81×10^{-4}
   b 14.9   **d** 9 080 000
2. **a** 10.6   **c** 0.03   **e** 0.002
   b 123.977   **d** 3.971   **f** 4.10
3. **a** 2000   **b** 2000   **c** 1990
4. 0.0004 (4 decimal places)

Estimation

1. **a** 2400   **b** 3
2. **a** A   **e** A   **i** B
   b C   **f** B   **j** A
   c B   **g** C
   d B   **h** A
3. **a** 6.7 – accept 6.5 to 6.9
   b 10.2 – accept 10.1 to 10.3
   c 12 – accept 10 to 14
   d 5 – accept 4 to 6

Upper and lower bounds

1. **a** 144.5 cm
   b 145.5 cm
   c $144.5 \leq l < 145.5$ cm

Review it!

1. **a** 24 647.515   **b** 21.5 (to 1 d.p.)
2. **a** 6.5   **c** 81
   b 4   **d** 5.94
3. **a** $5\frac{1}{3}$ **b** $9\frac{7}{13}$
4. $4\frac{3}{4} + \frac{1}{2}$
5. **a** lower bound = 10.113 cm² (to 3 d.p.)
       upper bound = 10.180 cm² (to 3 d.p.)
   b 10 cm²
6. **a** 3.34×10^{23} molecules
   b 2.99×10^{-26} kg
7. 0.16
8. **a** 1   **c** $\frac{1}{64}$
   b 3   **d** $\frac{1}{4}$
9. $5.55 \leq y < 5.65$
10. $\frac{1-\sqrt{2}}{1+\sqrt{2}} = \frac{1-\sqrt{2}}{1+\sqrt{2}} \times \frac{1-\sqrt{2}}{1-\sqrt{2}} = \frac{1-2\sqrt{2}+2}{1-2} = \frac{3-2\sqrt{2}}{-1} = 2\sqrt{2} - 3$
11. $111.5 \leq a < 112.5$
12. $\frac{8}{11}$
13. **a** $0.287996 \leq c < 0.289272$
   b 0.29 (to 2 s.f.)
14. **a** $\frac{2}{3}$
   b 80
15. **a** 4.5×10^{-7}
   b 1.2×10^{7}
   c 5.64×10^{3}
16. 3.2×10^{-1}
17. **a** 1, 2, 4, 8, 16, 32, 64   **b** 4

Algebra

Simple algebraic techniques

1. **a** formula   **c** expression   **e** equation
   b identity   **d** identity
2. **a** $10x^2 + 4x$ **c** $-3x^2 + 10xy$
   b $3a - b$ **d** $3x^3 - x - 5$
3. 16
4. 4

Removing brackets

1. **a** $2x + 8$ **c** $x - 1$ **e** $3x^2 + 3x$
   b $63x + 21$ **d** $3x^2 - x$ **f** $20x^2 - 8x$
2. **a** $5x + 12$ **c** $4x^2 + 2x$
   b $3x + 45$ **d** $3x^2 - 10x + 8$
3. **a** $p^{\frac{4}{3}}$ **b** r^{-1} **c** b^{-6}
4. **a** $t^2 + 8t + 15$ **c** $6y^2 + 41y + 63$
   b $x^2 - 9$ **d** $4x^2 - 4x + 1$
5. **a** $2x^3 + 21x^2 + 55x + 42$
   b $24x^3 - 46x^2 + 29x - 6$

Factorising

1. **a** $6(4t + 3)$ **c** $5y(x + 3z)$
   b $a(9 - 2b)$ **d** $6xy^2(4x^2 + 1)$
2. **a** $(x + 7)(x + 3)$ **c** $(2x + 5)(3x + 2)$
   b $(x + 5)(x - 3)$ **d** $(2x + 7)(2x - 7)$
   e $(5x + \sqrt{7})(5x - \sqrt{7})$
3. $\frac{2x - 1}{x + 3}$
4. $\frac{1}{2x + 3}$

Changing the subject of a formula

1. **a** $r = \sqrt{\frac{A}{\pi}}$ **b** $r = \sqrt{\frac{A}{4\pi}}$ **c** $r = \sqrt[3]{\frac{3V}{4\pi}}$
2. **a** $c = y - mx$ **d** $u = \sqrt{v^2 - 2as}$
   b $a = \frac{v - u}{t}$ **e** $t = \frac{2s}{u + v}$
   c $s = \frac{v^2}{2a}$ **f** $x = \frac{2pq}{2p + q}$

Solving linear equations

1. **a** $x = 3$ **b** $x = 3$ **c** $x = 20$
2. **a** $x = 5$ **b** $x = 18$ **c** $x = 20$
3. **a** $x = -2$ **b** $m = 1$ **c** $x = \frac{6}{5}$, $1\frac{1}{5}$ or 1.2

Solving quadratic equations using factorisation

1. **a** $x = -2$ or $x = -3$
   b $x = -3$ or $x = 4$
   c $x = -\frac{7}{2}$ or $x = -5$
2. **a** Area $= \frac{1}{2} \times$ base $\times$ height
        $\frac{1}{2}(2x + 3)(x + 4) = 9$
        $\frac{1}{2}(2x^2 + 11x + 12) = 9$
        $2x^2 + 11x + 12 = 18$
        $2x^2 + 11x - 6 = 0$
   b $x = \frac{1}{2}$ **c** base = 4 cm, height = 4.5 cm

3 By Pythagoras' theorem $(x + 1)^2 + (x + 8)^2 = 13^2$

$x^2 + 2x + 1 + x^2 + 16x + 64 = 169$

$2x^2 + 18x - 104 = 0$

Dividing through by 2 gives

$x^2 + 9x - 52 = 0$

$(x - 4)(x + 13) = 0$

So $x = 4$ or -13 (disregard $x = -13$ as x is a length)

Hence $x = 4$ cm

Solving quadratic equations using the formula

1 $x = 2.14$ or $x = -1.64$ (to 3 s.f.)

2 a $\frac{2x + 3}{x + 2} = 3x + 1$

$2x + 3 = (3x + 1)(x + 2)$

$2x + 3 = 3x^2 + 7x + 2$

$0 = 3x^2 + 5x - 1$

$3x^2 + 5x - 1 = 0$

b $x = -1.85$ or $x = 0.18$ (to 2 d.p.)

Solving simultaneous equations

1 a $x = 2, y = -1$

b $x = 4, y = 2$

2 $x = \frac{1}{5}, y = -2\frac{3}{5}$ or $x = \frac{1}{2}, y = -2$

3 Equating the y values gives

$x^2 + 5x - 4 = 6x + 2$

$x^2 - x - 6 = 0$

$(x - 3)(x + 2) = 0$

$x = 3$ or -2

When $x = 3, y = 6 \times 3 + 2 = 20$

When $x = -2, y = 6 \times (-2) + 2 = -10$

Points are $(3, 20)$ and $(-2, -10)$

Solving inequalities

1 a $x > 6$

$\{x: x > 6\}$

b $x \geq 11$

$\{x: x \geq 11\}$

c $x < 26$

$\{x: x < 26\}$

2 a $x > 10$ **b** $x < 0.4$ or $\frac{2}{5}$ **c** $x \leq 8$

3 a

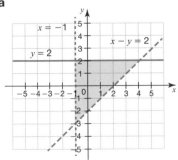

b $(0, 2), (0,1), (0, 0), (0, -1), (1, 2), (1, 1), (1, 0), (2, 2),$ $(2, 1), (3, 2)$

4 $x < -2$ and $x > 5$

Problem solving using algebra

1 26, 51   **2** 65   **3** 9 cm by 3 cm

Use of functions

1 a $\frac{x}{5} - 4$ **b** $5(x + 4)$

2 a i $\frac{1}{2}$ **ii** 4

b $\frac{2}{3}$ **c** $\frac{1}{x - 1}$

3 a -1 **b** $-\frac{2}{3}$ **c** $\frac{x + 1}{x}$

4 a $\sqrt{x^2 + 8x + 7}$ **c** 4

b $\sqrt{(x^2 - 9)} + 4$

Iterative methods

1 $x = 0.8$

2 1.521

Let $f(x) = x^3 - x - 2$

$f(1.5215) = (1.5215)^3 - 1.5215 - 2 = 0.0007151$

$f(1.5205) = (1.5205)^3 - 1.5205 - 2 = -0.005225$

As there is a change in sign, $a = 1.521$ to 3 decimal places.

Equation of a straight line

1 a 2   **b** $-\frac{1}{2}$ **c** $y = -\frac{1}{2}x + 5$

2 $y = 3x - 3$

3 $2x - y + 2 = 0$

4 a $\frac{1}{2}$ **b** $(2, 2)$ **c i** -2

ii $y = -2x + 6$

Quadratic graphs

1 a $2(x - 3)^2 - 17$ **c**

b i $(3, -17)$

ii $x = 0.1$ and $x = 5.9$

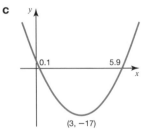

2 a $y = x^2 - 4x - 5$ **b** $y = -x^2 + 9x - 14$

3 a $x^2 + 12x - 16 = (x + 6)^2 - 52$

b $(-6, -52)$

Recognising and sketching graphs of functions

1 a B   **c** E   **e** D

b F   **d** A   **f** C

2 a

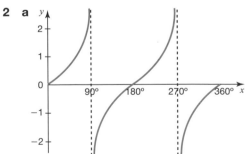

b $x = 240°$

3 a A   **b** G   **c** F   **d** E

Translations and reflections of functions

1 a $(3, 5)$ **c** $(2, -5)$

   b $(-1, 5)$ **d** $(-2, 5)$

2 a

  b

Equation of a circle and tangent to a circle

1 a $(0, 0)$ **b** 7

2 a $x^2 + y^2 = 100$

   b Gradient of radius to $(8, 6) = \frac{6}{8} = \frac{3}{4}$

      Gradient of tangent $= -\frac{4}{3}$

   c $y = -\frac{4}{3}x + 16\frac{2}{3}$

Real-life graphs

1 a $5\,\text{km/h}$ **b** 0.25 hours   **c** $24\,\text{km/h}$

2 a

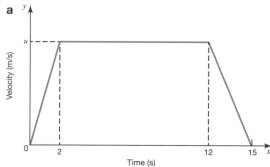

   b $u = 4\,\text{m/s}$ **c** $1.33\,\text{m/s}^2$

Generating sequences

1 a 17 **c** -12 **e** $\frac{1}{48}$

   b 3.0 **d** 432 **f** $-\frac{1}{16}$

2 $17, 290$

3 $2.25, 5.5$

The nth term

1 a $47, 44, 41$

   b no   **c** -1

2 a $6, 18, 54, 162$

   b Both 2 and 3 are factors, so 6 must also be a factor.

3 a nth term $= \frac{1}{n}$

   b nth term $= \frac{1}{(\sqrt{2})^n}$ or $\left(\frac{1}{\sqrt{2}}\right)^n$

4 a nth term $= 2n - 3$ **b** $x = 31$

5 nth term $= 4n^2 + n - 1$

Arguments and proofs

1 a True: $2n$ is always even as it is a factor of 2. Adding 1 to an even number always gives an odd number.

   b False: $x^2 = 9$, so $x = \sqrt{9} = \pm 3$.

   c False: n could be a decimal such as 4.25, so squaring it would not give an integer.

   d False: if $n \leq 1$ this is not true.

2 Let the four consecutive numbers be n, $n + 1$, $n + 2$ and $n + 3$.

   Sum $= n + (n + 1) + (n + 2) + (n + 3)$

           $= 4n + 6$

           $= 2(2n + 3)$

   Since the sum is a multiple of 2, it is always even.

   Therefore the sum of four consecutive numbers is always even.

3 Let the consecutive integers be x, $x + 1$ and $x + 2$.

   Sum of the integers $= x + x + 1 + x + 2 = 3x + 3$

   $= 3(x + 1)$

   As 3 is a factor, the sum must be a multiple of 3.

4 a The numerator is larger than the denominator so the fraction will always be greater than 1. Statement is false.

   b As a is larger than b, squaring a will result in a larger number than squaring b. Hence $a^2 > b^2$ so the statement is false.

   c The square root of a number can have two values one positive and the other negative so this statement is false.

Review it!

1 a $-9x + 12$ **b** $6x + 4$ **c** $6x^3 + 25x^2 + 16x - 15$

2 a $(2x - 1)(x + 4)$ **b** $x = \frac{1}{2}$ or $x = -4$

3 a $8x^6y^3$ **b** $6x$ **c** $\frac{5}{b}$

4 $x = 2, y = 1$

5 a $\dfrac{3}{x + 7} = \dfrac{2 - x}{x + 1}$

     $3(x + 1) = (2 - x)(x + 7)$

     $3x + 3 = 2x + 14 - x^2 - 7x$

     $3x + 3 = -x^2 - 5x + 14$

     $x^2 + 8x - 11 = 0$

   b $x = 1.20$ or $x = -9.20$ (to 2 d.p.)

6 $x = \dfrac{3y - 2z}{az + 1}$

7 a $f^{-1}(x) = 3x - 15$ or $f^{-1}(x) = 3(x - 5)$

   b $k = 7$

8 a $26, 22, 18$ **b** 8th term $= -2$

9 The point lies outside the circle.

10 $x - 9y$

11 a $2(x + 2)^2 - 7$

   b i $(-2, -7)$

      ii $x = -3.9$ and $x = -0.1$

   c

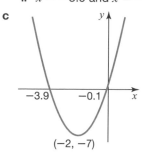

12 $30\,\text{cm}^2$

13 $y = 7x + 12$

14 $x = -2, y = 0$ or $x = \frac{6}{5}, y = \frac{8}{5}$

Ratio, proportion and rates of change

Introduction to ratios

1 **a** $1:3$ **b** $5:12$ **c** $4:9$

2 **a** $1:8$ **b** $100:1$ **c** $1:50$

3 £250, £150

4 315 members

5 £42 000

6 62.9%

7 £5.70

8 $10:5:1$

Scale diagrams and maps

1 30 cm

2 **a** $1:500\,000$ **b** 6 km

Percentage problems

1 2.67% **3** £39 330 **5** £2520

2 88.3% **4** £356

Direct and inverse proportion

1 Inverse proportion means that if one quantity doubles the other quantity halves.

2 **a** $y = kx$ **b** 10.7

3 **a** cheaper in the UK

b £5.49 cheaper in the UK

4 $268\,\text{cm}^3$

5 33 333 (to nearest whole number)

6 150

7 **a** $A = 6x^2, k = 6$ **b** $96\,\text{cm}^2$

Graphs of direct and inverse proportion and rates of change

1 Graph C

2 Graph B

3 $a = 6$

4 Equation connecting x and y is $y = \frac{k}{x}$

When $x = 1, y = 4$ so $4 = \frac{k}{1}$ so $k = 4$

The equation of the curve is now $y = \frac{4}{x}$

When $x = 4, y = \frac{4}{4} = 1$ so $a = 1$

When $y = 0.8, 0.8 = \frac{4}{x}$ giving $x = 5$ so $b = 5$

Hence $a = 1$ and $b = 5$

5 $y = kx^2$

When $x = 2, y = 16$ so $16 = k \times 2^2$ giving $k = 4$

$y = 4x^2$

Hence $36 = 4x^2$ so $x = 3$

$a = 3$

Growth and decay

1 **a** 1.05 **c** 1.0375

b 1.25 **d** 0.79

2 £4962 (to nearest whole number)

3 £745.86 (to nearest penny)

4 2048

Ratios of lengths, areas and volumes

1 $6\frac{2}{3}\,\text{cm}$

2 $\frac{V_A}{V_B} = \frac{27}{64} = (\text{scale factor})^3$, so scale factor $= \frac{3}{4}$

$\frac{A_A}{A_B} = (\text{scale factor})^2 = \frac{9}{16}$

So $A_A = \frac{9}{16} \times 96 = 54\,\text{cm}^2$

3 **a** Triangles ABE and ACD must be proved similar:

BE parallel to CD, all the corresponding angles in both triangles are the same.

$BE = 4\,\text{cm}$

b $10\,\text{cm}^2$

Gradient of a curve and rate of change

1 **a** The gradient represents the acceleration.

b $5\,\text{m/s}^2$

c 3.9 s

Converting units of areas and volumes, and compound units

1 **a** **i** $14\,800\,\text{mm}^2$

ii $0.0148\,\text{m}^2$

b $0.000\,12\,\text{m}^3$

2 $1.932\,\text{g/cm}^3$ (to 3 d.p.)

3 2 000 000

4 $13.89\,\text{m/s}$ (to 2 d.p.)

5 **a** 56 km/h

b The distance will not be the same, so the average speed will be different.

Review it!

1 £96

2 **a** $5.10\,\text{m}^2$

b $5.31\,\text{m}^2$

3 $y = 2$

4 243 students

5 £485 000

6 **a** £2250

b £2262.82

7 **a** 3420 yuan

b travel agent

8 £30 000

9 $0.64\,\text{cm}^3$

10 20%

11 $8.7\,\text{g/cm}^3$

Geometry and measures

2D shapes

1 a true   **c** true   **e** true
   b false   **d** false   **f** true

Constructions and loci

1

locus of point X

2 This is a reduced version of the answer.

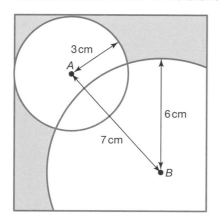

Properties of angles

1 144°

2 a 9   **b** 1260°

3 20°

4 a x = angle EBC = 55° (alternate angles)

   b Angle EHI = 180 − (85 + 55) = 40° (angle sum in a triangle)

     Angle DEH = angle EHI = 40° (alternate angles)

5 77°, 103°, 77°, 103° (a = 24).

Congruent triangles

1 $AC = AC$ (common to both triangles)
$AB = AD$ (given)
$BC = CD$ (given)
Triangles ACD and ACB are congruent (SSS).
Hence angle ABC = angle ADC

2 $AB = AC$ (given)
$BM = MC$ (M is the midpoint of BC)
$AM = AM$ (common to both triangles)
Triangles ABM and ACM are congruent (SSS)
Hence angle AMB = angle AMC
Angle AMB + angle AMC = 180°, so angle
$AMB = \frac{180}{2} = 90°$

Transformations

1 a, b

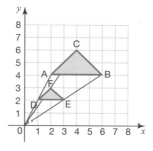

   c Enlargement, scale factor 2, centre of enlargement (0, 0)

2

3

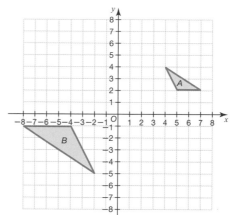

4 Translation by $\begin{pmatrix} 5 \\ -4 \end{pmatrix}$

Invariance and combined transformations

1 a, b, c

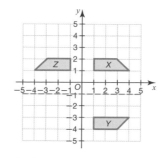

   d Reflection in the y-axis

2 a, b

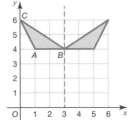

c (3, 4)

3 a, b

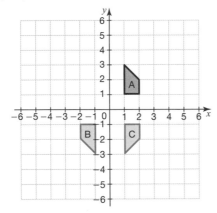

c A reflection in the *x*-axis.

3D shapes

1

Shape	Number of vertices	Number of faces	Number of edges
Triangular-based pyramid	4	4	6
Cone	1	2	1
Cuboid	8	6	12
Hexagonal prism	12	8	18

2 $V = 16, F = 10, E = 24; V + F - E = 16 + 10 - 24 = 2$

Parts of a circle

1 a radius **c** minor arc **e** minor sector

b chord **d** minor segment

Circle theorems

1 a angle ACB = angle YAB = 30° (alternate segment theorem)

b angle ABC = 90° (angle in a semicircle)

c angle ADC = 90° (angle in a semicircle)

2 a angle OAX = 90° (angle between tangent and radius)

b angle AOX = 180 − (90 + 30) = 60° (angle sum in a triangle)

c angle ACB = 30° (angle at centre twice angle at circumference)

3 a angle ADB = 40° (angles on same arc)

b angle EDB = angle EDA + angle ADB = 50 + 40 = 90°

So BD is a diameter of the circle (angle between tangent and diameter is 90°)

angle BAD = 90° (angle in a semicircle)

4 Angle ACB = 46° (angles bounded by the same chord in the same segment are equal)

Angle ABC = 90° (angle in a semi-circle is a right-angle)

Angle BAC = 180 − (90 + 46) = 44° (angles in a triangle add up to 180°)

5 a angle ABC = 60° (angle in an equilateral triangle)

angle ADC = 180 − 60 = 120° (opposite angles in a cyclic quadrilateral add up to 180°)

b angle BAC = 60° (angle in an equilateral triangle)

angle CBQ = angle BAC (alternate segment theorem) = 60°

Projections

1 a **c**

b

2

Bearings

1 245°

2 a 225° **b** 320°

Pythagoras' theorem

1 a 10.3 cm (to 1 d.p.) **b** 8.8 cm (to 1 d.p.)

2 height = 10.91 cm

area = 54.54 cm²

3 x = 8.12 cm (to 2 d.p.)

4 Let the perpendicular height of the triangle = h

Area of triangle = $\frac{1}{2} \times 14 \times h$

Hence $\frac{1}{2} \times 14 \times h = 90$

Solving gives h = 12.8571 cm

By Pythagoras' theorem, $AC^2 = 7^2 + 12.8571^2$

Solving gives AC = 14.6 cm (3 significant figures)

Area of 2D shapes

1 £1596

2 $\frac{5}{8}$

3 a area of semicircle = $\frac{1}{2} \times \pi \times x^2 = \frac{\pi x^2}{2}$

area of rectangle = $4x \times 2x = 8x^2$

area of shape = $\frac{\pi x^2}{2} + 8x^2 = x^2\left(8 + \frac{\pi}{2}\right)$

b $x(10 + \pi)$

Volume and surface area of 3D shapes

1 $r = \sqrt{\frac{3}{2}a^3}$

2 **a** $12.5\,\text{m}^2$ **b** $62.5\,\text{m}^3$ **c** 10 hours (to nearest hour)

3 $1200\,\text{m}^2$ (to 3 s.f.)

4 $109.5\pi\,\text{cm}^3$

Trigonometric ratios

1 **a** 17.32 cm (to 2 d.p.)

   b 9.19 cm (to 2 d.p.)

2 **a** 73.3° (to 0.1°)

   b 50.3° (to 0.1°)

3 $\sin\theta = \frac{b}{c}$ and $\cos\theta = \frac{a}{c}$

   $\frac{\sin\theta}{\cos\theta} = \frac{\frac{b}{c}}{\frac{a}{c}} = \frac{b}{c} \times \frac{c}{a} = \frac{b}{a} = \tan\theta$

4 **a** 13.60 cm (to 2 d.p.)

   b 63.8° (to 1 d.p.)

   c 8.56 cm (to 2 d.p.)

Exact values of sin, cos and tan

1 $\sqrt{3}\tan 30° + \cos 60° = \sqrt{3} \times \frac{1}{\sqrt{3}} + \frac{1}{2} = 1 + \frac{1}{2} = \frac{3}{2}$

   $a = 3, b = 2$

2 $\sin 30 = \frac{1}{2}$ and $\cos 30 = \frac{\sqrt{3}}{2}$

   $\sin^2 30 = \frac{1}{4}$ and $\cos^2 30 = \frac{3}{4}$

   $\sin^2 30 + \cos^2 30 = \frac{1}{4} + \frac{3}{4} = 1$

Sectors of circles

1 47.7° (to 1 d.p.)

2 **a** 7.99 cm **b** 3.94 cm²

3 Radius $OA = 10.3\,\text{cm}$

4 Area of sector = $109\,\text{cm}^2$

Sine and cosine rules

1 **a** 15 cm²

   b 15.5 cm (to 3 s.f.)

2 27.7° (to 0.1°)

3 **a** 38.6°

   b If angle XZY is not obtuse, then angle XYZ can be obtuse. An alternative answer would be 141.4°.

Vectors

1 **a** $\begin{pmatrix} -6 \\ 8 \end{pmatrix}$ **b** $\begin{pmatrix} 2 \\ 0 \end{pmatrix}$

2 **a** $\mathbf{b} - \mathbf{a}$ **b** $\frac{3}{5}(\mathbf{b} - \mathbf{a})$ **c** $\frac{1}{5}(2\mathbf{a} + 3\mathbf{b})$

3 **a** $-\mathbf{a} - 3\mathbf{b}$

   b $\overrightarrow{BM} = \frac{1}{2}\overrightarrow{BC} = \frac{1}{2}(-\mathbf{a} - 3\mathbf{b})$

   $\overrightarrow{PM} = \overrightarrow{PA} + \overrightarrow{AB} + \overrightarrow{BM} = 2\mathbf{b} + \mathbf{a} + \frac{1}{2}(-\mathbf{a} - 3\mathbf{b})$

   $= \frac{1}{2}\mathbf{b} + \frac{1}{2}\mathbf{a} = \frac{1}{2}(\mathbf{a} + \mathbf{b})$

   $\overrightarrow{MD} = \overrightarrow{MB} + \overrightarrow{BD} = -\frac{1}{2}(-\mathbf{a} - 3\mathbf{b}) + \mathbf{a} = \frac{3}{2}\mathbf{a} + \frac{3}{2}\mathbf{b}$

   $= \frac{3}{2}(\mathbf{a} + \mathbf{b})$

$\overrightarrow{PM}$ and $\overrightarrow{MD}$ have the same vector part $(\mathbf{a} + \mathbf{b})$, therefore they are parallel. Both lines pass through M and parallel lines cannot pass through the same point unless they are the same line. Hence *PMD* is a straight line.

Review it!

1 Angle $ADC = (180 - x)°$ (opposite angles of cyclic quadrilateral)

   Angle $ADE = y°$ (alternate segment theorem)

   Angle $CDF = 180 -$ angle $ADE -$ angle ADC
   $= 180 - y - (180 - x) = (x - y)°$

2

$\sin 60° = \frac{\sqrt{3}}{2}$ $\cos 45° = \frac{1}{\sqrt{2}}$

$\cos 45° + \sin 60° = \frac{1}{\sqrt{2}} + \frac{\sqrt{3}}{2}$

$= \frac{1}{\sqrt{2}} \times \frac{\sqrt{2}}{\sqrt{2}} + \frac{\sqrt{3}}{2}$

$= \frac{\sqrt{2}}{2} + \frac{\sqrt{3}}{2}$

$= \frac{1}{2}(\sqrt{2} + \sqrt{3})$

3 5, 12 and 13

4 **a**

   b

5 **a**

   b **i**

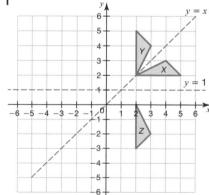

   b **ii** (2, 2)

   c See triangle Z on grid above

   d Rotation of 90° clockwise about (1, 1)

6 **a** 19.4 km (to 3 s.f.)

   b 280°

Probability

The basics of probability

1 $\frac{8}{125}$

2 a 0   **b** $\frac{1}{5}$ **c** $\frac{11}{20}$

Probability experiments

1 a 0.21 (to 2 d.p.)

  b 0.14 (to 2 d.p.)

  c Sean is wrong. 120 spins is a small number of spins and it is only over a very large number of spins that the relative frequencies may start to be nearly the same.

2 a 0.04   **b** 600 cans

3 45 apples

The AND and OR rules

1 a Independent events are events where the probability of one event does not influence the probability of another event occurring. Here it means that the probability of the first set of traffic lights being red does not affect the probability of the second set being red.

  b 0.06   **c** 0.56

2 a $\frac{7}{48}$ **b** $\frac{1}{32}$

Tree diagrams

1 a $\frac{1}{15}$ **b** $= \frac{7}{15}$

2 20 balls

3 a

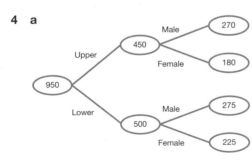

  b 0.15   **c** 0.85

4 a

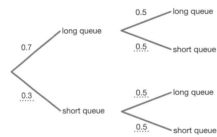

  b $\frac{109}{190}$ or 0.57

Venn diagrams and probability

1 a **i** {1, 3, 4, 5, 8, 9, 10, 11}

    ii {8, 9}

    iii {2, 5, 6, 10, 13}

  b $\frac{7}{11}$

2

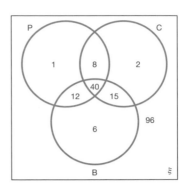

  a $\frac{10}{21}$ **b** $\frac{3}{28}$ **c** $\frac{48}{61}$

3 a

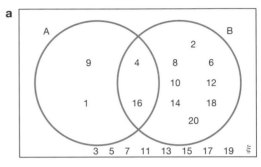

  b $P = \frac{8}{20} = \frac{2}{5}$

4 a $\frac{1}{12}$ **b** $\frac{5}{24}$

5 a 710   **b** $P(B|A) = \dfrac{P(A \text{ and } B)}{P(A)}$

$$P(B) = \frac{165 + 74}{710}$$

$$= \frac{239}{710}$$

$$= 0.34$$

$$P(B \text{ given } A) = \frac{165}{400}$$

$$= 0.41$$

As P(B given A) ≠ P(B) the two events are not independent.

6 a $\frac{3}{7}$ **b** $\frac{3}{5}$

Review it!

1 a

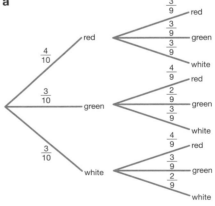

  b $\frac{4}{15}$

  c $\frac{11}{15}$

2 a

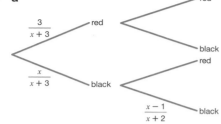

P(two black) $= \left(\frac{x}{x+3}\right) \times \left(\frac{x-1}{x+2}\right) = \frac{7}{15}$

$15x(x-1) = 7(x+3)(x+2)$

$15x^2 - 15x = 7x^2 + 35x + 42$

$8x^2 - 50x - 42 = 0$

$4x^2 - 25x - 21 = 0$

b 10 balls

c $\frac{7}{15}$

3

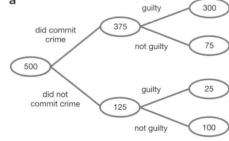

a $\frac{11}{60}$ **b** $\frac{15}{29}$

4 a

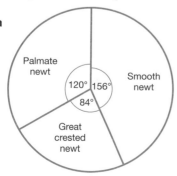

b $\frac{13}{20}$ **c** $\frac{1}{5}$

Statistics

Sampling

1 26

Two-way tables and pie charts

1 a

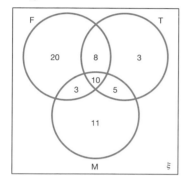

b No. The other pond might have had more newts in total. The proportion of smooth newts in the second pond is lower, but there may be more newts.

Line graphs for time series data

1 The trend is decreasing sales.

2 a

b The percentage of people using the local shop is decreasing.

c 44%

d There are no points so you would be trying to predict the future. There may be a change of ownership or a refurbishment making it more popular. It could even close down before then.

Averages and spread

1 13.5 years

2 Mean $= \frac{\text{total number of marks}}{\text{number of students}}$, so

Total mark for boys $= 50 \times 10 = 500$

Total mark for girls $= 62 \times 15 = 930$

Total mark for class $= 500 + 930 = 1430$

Mean for class $= \frac{\text{total number of marks}}{\text{number of students}} = \frac{1430}{25} = 57.2\%$

Joshua is wrong, because he didn't take account of the fact that there was a different number of boys and girls.

3 a

Cost (£C)	Frequency	Mid-interval value	Frequency × mid-interval value
$0 < C \le 4$	12	2	24
$4 < C \le 8$	8	6	48
$8 < C \le 12$	10	10	100
$12 < C \le 16$	5	14	70
$16 < C \le 20$	2	18	36

b £7.51 (to nearest penny)

Histograms

1

Mass (m pounds)	Frequency
$0.0 < m \le 0.5$	30
$0.5 < m \le 1.0$	**16**
$1.0 < m \le 1.5$	13
$1.5 < m \le 2.0$	21
$2.0 < m \le 2.5$	**32**
$2.5 < m \le 3.0$	38

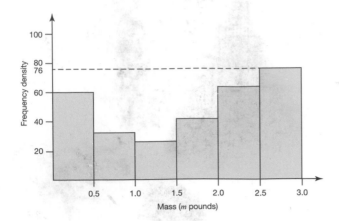

Cumulative frequency graphs

1 a

b 13 minutes

c **i** 17.5 minutes

ii 9 minutes

iii 8.5 minutes

d 5.6% (to 1 d.p.)

Comparing sets of data

1

	Height (cm)
Lowest height	38
Lower quartile	52
Median	57
Upper quartile	59
Highest height	69

2 a **i** 120 marks

ii 65 marks

iii 75 marks

iv 51 marks

v 24 marks

b For the girls: median mark = 74 marks, upper quartile = 89 marks, lower quartile = 58 marks and interquartile range = 31 marks

Comparison:

The median mark for the girls is higher (or higher average mark).

The interquartile range is lower for the boys showing that their marks are less spread out for the middle half of the marks.

Scatter graphs

1 a, b

c 71 visitors

d **i** 120 visitors

ii There are no points near this temperature so you cannot assume the trend continues.

Review it!

1 a 36 students   **c** $4 < a \leq 6$

b £5.83 (to the nearest penny)

2 a **i**

b 54   **c** 40

3 a D (the median mark is furthest to the right)

b C (the largest gap between the quartiles)

c D as the median mark is the highest and also the interquartile range is small which means 50% of pupils got near to the median mark.

4 a The sample needs to be representative of people living on the street: male/female, adults/children.

b 120 people